BERING STRAIT: The Regional Physical Oceanography

BERING STRAIT

The Regional Physical Oceanography

L. K. COACHMAN, K. AAGAARD, R. B. TRIPP

UNIVERSITY OF WASHINGTON PRESS
Seattle and London

Copyright © 1975 by the University of Washington Press
Printed in the United States of America

All rights reserved. No part of this publication may be
reproduced or transmitted in any form or by any means,
electronic or mechanical, including photocopying, recording,
or any information storage or retrieval system, without
permission in writing from the publisher.

Library of Congress Cataloging in Publication Data
Coachman, L   K   1926–
  Bering Strait: the regional physical oceanography.
  Bibliography: p.
  Includes index.
  1. Oceanography—Bering Strait. I. Aagaard,
Knut, 1939–   joint author. II. Tripp,
Richard B., joint author. III. Title.
GC411.C62   551.4′65′5   75-40881
ISBN 0-295-95442-6

Affectionately dedicated to

CLIFFORD ADRIAN BARNES

whose careful scientific judgment, immense integrity, and human kindness are standards of excellence in professional and personal conduct

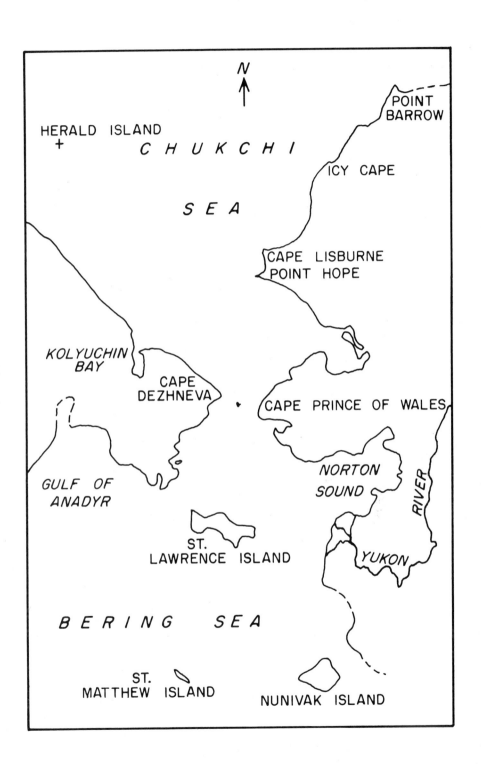

(*Overleaf*) A combined visual-infrared photograph of Bering Strait taken 22 May 1974 from the NOAA-3 satellite. There was little cloud cover obscuring the surface features on this unusually clear day, only some very thin clouds east and south of St. Lawrence Island and over the Gulf of Anadyr and some heavier patches near Herald Island.

Smooth, unbroken areas of ice (shore-fast ice?) can be clearly distinguished from areas with broken ice which are associated in part with a more vigorous water motion. The current was directed due north from Bering Strait, being steered by the Cape Prince of Wales shoal before turning eastward toward Kotzebue Sound. It then curved toward the north again and closed the coast near Kivalina (SE of Pt. Hope). Note the areas of smooth ice in the lee of Cape Lisburne and Icy Cape and the narrow band of ice along the north shore of St. Lawrence Island.

In infrared photographs an index to surface temperature variations is afforded by the degree of shading between white (coldest) and black (warmest). Areas which were notably free of ice, e.g., Norton Sound, the large area extending northward from the Strait of Anadyr, an area south of St. Lawrence Island, and northeast from Cape Lisburne along the shore-fast ice, are also dark, suggesting a connection between areas of accumulated or advected heat and centers of ice disintegration. Note the arc-shaped area of broken ice extending well north from Bering Strait.

Local winds also distribute loose ice, even contrary to the currents, which probably accounts for the band of broken ice between Bering Strait and the north shore of St. Lawrence Island. Winds may also account for the band of ice extending southwest from St. Lawrence. However, this area is the so-called "cold center" of the Bering Sea shelf, an area of weak flow. The ice band may therefore instead be a winter residue unaffected by currents.

(*Left*) An infrared photograph of Bering Strait taken 22 October 1974 from the NOAA-3 satellite. The temperature range covered by the degree of shading is from approximately — 7.8°C (white) to +6.1°C (black).

Heavy cloud bands obscured the southern and northern portions; one cloud boundary ran southwest across the middle of Norton Sound and eastern St. Lawrence Island, and another from Cape Lisburne to Kolyuchin Bay. The Diomedes, King, and Sledge islands are clearly visible, and there was as yet no ice in the area.

The fascinating feature is the distribution of surface temperatures. Norton Sound and the Gulf of Anadyr were relatively warm, while Kotzebue Sound had experienced a considerable autumnal heat loss. A contiguous warm band extending northward through Bering Strait is suggestive of the dominant advection, and extended due north from Cape Prince of Wales (compare photo overleaf). However, the sharp bend to the west in the southern Chukchi Sea is probably not representative of the current field. The temperature indicated by the infrared photography is that of a very thin surface layer only, and a change in the atmospheric exchange and the vertical mixing regime can produce a cooler surface temperature even though the heat content of the subsurface water remains high. We believe the circulation pattern was probably the normal one, curving eastward toward Kotzebue and then northwestward south of Pt. Hope.

(Photos courtesy of R. D. Muench, University of Alaska)

"In the month of September on the 20th day [1648], en route from the Kovyma [Kolyma] River by sea, at a harbor the trader Fedot Alekseyev [Popov] was wounded by the Chukchi people in a fight, and that same Fedot, while he was with me, Semeika, the sea carried away without leaving a trace, and carried me, Semeika, involuntarily in all directions until after Pokrova Bogoroditsi [a feast day, Oct. 1 (14)], and threw me on the shore on a headland beyond the Anandyr [Anadyr] River . . . .

"However, from the Kovyma River it is possible to go by sea towards the Anandyr River, and there is a nose [Cape Dezhneva], which extends far into the sea . . . and opposite to the nose there are two islands [Diomedes], and on these islands live Chukchi [actually Eskimos] . . . ."

S. Dezhnev, Report to the Commander of the Army in Yakutsk, April 1655.

"On the 15th of August [1728] we arrived in the latitude of 67°18' and I judged that we had clearly and fully carried out the instructions given by his Imperial Majesty of glorious and ever deserving memory, because the land no longer extended to the north. Neither from the Chukchi coast nor to the eastward could any extension of the land be observed."

Report of Fleet-Captain Bering on his Expedition to the Eastern Coast of Siberia.

ACKNOWLEDGMENTS

WE ARE INDEBTED to the many people who supported the collection of data. Drs. Max Britton and Ned Ostenso and Mr. Ron McGregor, all of the Office of Naval Research, gave freely of their advice and encouragement; the officers and men of the United States Coast Guard provided field assistance on numerous expeditions aboard the *Staten Island* and the *Northwind*; and Dr. Max Brewer and Mr. John Schindler, former directors of the Naval Arctic Research Laboratory, Barrow, Alaska, provided logistical support for many of the winter expeditions. Drs. Clifford Barnes, George Pickard, and Joe Reid reviewed the manuscript in a particularly helpful way.

The work was done primarily under ONR contracts N-00014-67A-0103-0014, N-00014-67A-0103-0021 and NSF grants GA 11147 and DES 74-14662.

CONTENTS

1. INTRODUCTION	3
Geographic Features of the Region	4
Investigations	6
2. WATER MASSES	11
Principal Water Masses	12
Water Mass Sources	17
Long-Term Temporal Variations	44
Short-Term Variations	59
Anomalous Events	66
3. CURRENTS	74
General Flow Field	74
Fluctuations in the Current Field	76
Transport	98
4. CHUKCHI SEA	111
Introduction	111
Water Masses	114
Currents	140
5. NUMERICAL CONSIDERATIONS	146
Dynamical Calculations	146
Eddy Coefficients	152
Heat Budget	154
6. SUMMARY AND CONCLUSIONS	157
Water Masses	158
Currents	160
Chukchi Sea	161
Numerical Considerations	162
REFERENCES	165
INDEX	169

BERING STRAIT: The Regional Physical Oceanography

1

INTRODUCTION

THE FLOW REGIME and property distributions in the northern Bering and Chukchi seas, connected by Bering Strait, are dominated by the general northward flow. That such a flow exists in this flat and shallow (~50 m) region has been known since the observations of Captain Vitus Bering in 1728; but the tools and techniques for accurate and systematic oceanographic study have only become available during recent decades.

All available observations in Bering Strait up to 1964 were analyzed by Coachman and Aagaard (1966), and the water properties and flow in the strait in summer were elucidated. The water properties in Bering Strait, based primarily on the detailed section obtained in 1964, were described as follows. In the eastern part, a sharp pycnocline at 10 to 15 m typically separates a surface layer of warm (6 to 10°C), low salinity (30 to 31°/$_{00}$) water, from colder (1 to 4°C), more saline (31 to 32.7°/$_{00}$) water, while the western part contains relatively uniformly cold (1 to 4°C) and saline (32.7 to 33°/$_{00}$) water. Relatively strong velocity shears occur between these water masses: the upper flow in the eastern part of the strait is typically quite swift, so that speeds >100 cm sec^{-1} are common; the deeper-lying water in the eastern part has intermediate speeds of 30 to 60 cm sec^{-1}; and the flow through the western channel is even slower and more uniform.

Everywhere in the strait the summer flow is nearly always northerly, transporting 1–2 Sv of Bering Sea water into the Arctic Ocean. Investigation of the dynamics of the system showed the primary driving mechanism to be a sea surface sloping downward to the north, the cause of which must probably be sought in the regional distribution of wind stress.

Coachman's and Aagaard's study was limited to the summer season in Bering Strait proper, leaving unanswered many questions concerning the physical oceanographic regime. What are the sources of the water masses found in the area? What are the seasonal modifications of the water masses? What causes the very large accelerations in the northward flow that appear to be particularly pronounced in the eastern channel? A good understanding of the Bering Strait hydrography must obviously include observations and analyses extending both over a long time and over the entire region of the northern Bering and southern Chukchi seas.

GEOGRAPHIC FEATURES OF THE REGION

Bering Strait presents a marked constriction to the flow; the land masses are more widely separated to the north in the Chukchi Sea and to the south in the northern Bering Sea (see Fig. 1) The whole region is uniformly shallow, and bottom slopes are very slight except on approach to land. There is a gentle depth gradient across the region, the eastern part, averaging between 30 to 40 m depth (outside of the major embayments), being the shallower, contrasted with 40 to 50 m on the western side.

There are two major embayments on the Alaskan shore: Norton Sound south of Bering Strait, and Kotzebue Sound to the north. On the Siberian shore, the Gulf of Anadyr west of St. Lawrence Island is relatively deep (70 to 90 m on the average); and a channel over 50 m deep extends north through the strait between the Chukotski peninsula of Siberia and St. Lawrence Island, which we shall call the Strait of Anadyr. Other notable channels (or depressions in the bottom) include a narrow channel more than 40 m deep rounding the eastern end of St. Lawrence Island; a channel deeper than 30 m extending eastward into Norton Sound close off Nome; and the terminus of the Hope Submarine Valley with depths greater than 60 m southwest of Pt. Hope (Creager and McManus 1966).

The land mass of St. Lawrence Island dominates the northern Bering Sea. It constrains a generally northward flow to either side and produces a lee effect noticeable at least 150 km to the north. It may reflect long waves and thus add to the complexity of the tides in the region; and it also gives rise to other oceanographic phenomena associated with land boundaries, such as upwelling and edge waves.

Five other islands in the region are relatively small: Sledge Island about 37 km west of Nome, King Island about 55 km WSW of Port Clarence, and Fairway Rock, Little Diomede, and Big Diomede (Ratmanov) islands all in Bering Strait. These all rise abruptly from the sea floor.

Two spitlike shoals occur in the region, and they are probably the result of the prevailing flow. One extends over 100 km north from Cape Prince of Wales, and the other about 18 km NW from Pt. Hope (Creager and McManus 1966).

Significant quantities of fresh water are introduced into the region between May/June and August, primarily from the southeast where the

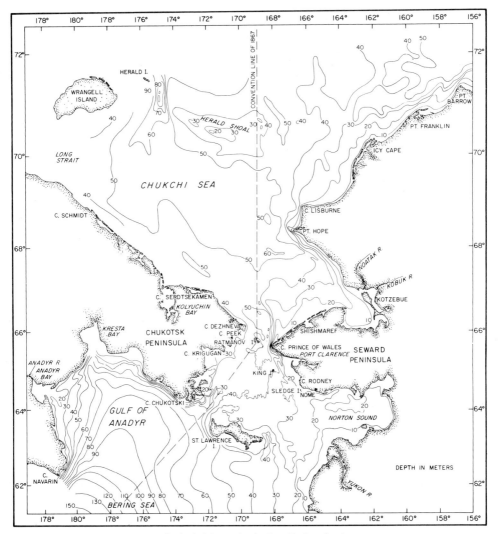

1. The northern Bering and Chukchi seas including Bering Strait

Yukon River empties into the south side of Norton Sound. Small but locally significant quantities of fresh water enter in summer into Kotzebue Sound from the Kobuk and Noatak rivers and into the western Gulf of Anadyr from the Anadyr River.

One further geographic feature merits comment. Many of the prominent capes in the region are high and mountainous, and winds tend to be accelerated locally to seaward of these points. Thus Bering Strait, which is only about 85 km wide, has the bold headlands of Cape Dezhneva on the west and Cape Prince of Wales on the east, and the winds in the strait locally tend to be accelerated and channeled in either northerly or southerly directions.

INVESTIGATIONS

Data available from Bering Strait through 1964 were surveyed by Coachman and Aagaard (1966). Some data from the surrounding waters (the Gulf of Anadyr, Bering Sea north of St. Lawrence, and the southern Chukchi Sea) were available at that time but were not included in the 1966 review. Since then, a number of investigations have been made, and all major surveys from the region through 1973 are summarized in table 1.

Some recent surveys (*Northwind* 1967; *Staten Island* 1968) were designed according to the criteria of Coachman and Aagaard (1966):

(1) Observations of currents and water properties must be made in considerable detail in both the vertical and horizontal extent.

(2) Observations must be repeated to measure accelerations and changes in water properties, both on short term and ultimately throughout the year.

(3) Observations must extend north and south of the strait to permit assessment of meridional variations.

The 1967 *Northwind* and 1968 *Staten Island* cruises are depicted in Figs. 2a, 2b, and 3. At each station the vessel was anchored, and a water bottle cast and vertical profile of currents with a deck read-out current meter were made. The stations were closely spaced across the system (to within 22 km [12 n. mi.] of the USSR), and the station lines were occupied from south to north to minimize advective changes in the observed scalar fields. The 1969 *Northwind* cruise duplicated the 1968 *Staten Island* stations, but without current measurements. The 1967 *Thompson* cruise conducted, within one week, four reoccupations of the Bering Strait section first occupied in 1964 by the *Northwind*. This section has been subsequently included in all summer cruises. The 1972 *Oshoro Maru* cruise began with the Bering Strait section and extended the coverage north to a section from Herald Island-Pt. Barrow.

The most extensive coverage of the Gulf of Anadyr has been that of the *Northwind* in 1962, which was repeated in part in 1970. In addition, during the 1970 cruise five current stations were occupied at which near-bottom measurements were made for about 30 hours each.

Ice conditions have severely restricted work in winter. However, the February 1968 *Northwind* cruise accomplished numerous hydrographic

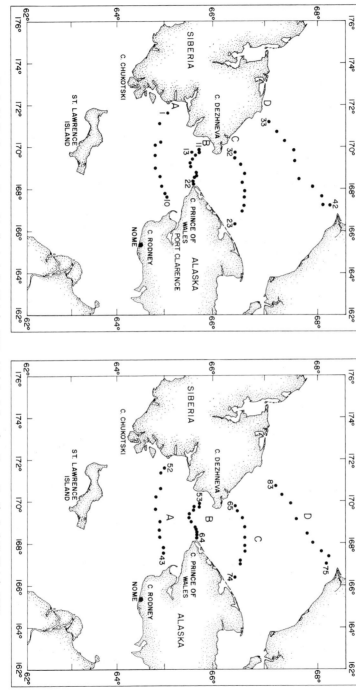

2. (Left) Station locations *Northwind* 12–17 July 1967; (right) station locations *Northwind* 18–23 July 1967.

TABLE 1
Oceanographic Surveys, Bering Strait Region

Year	Dates	Vessel	Region	Measurements
1922	26 Jul–5 Oct (Sverdrup 1929)	*Maud*	Chukchi Sea	15 hydro stations
1932	29 Jul–8 Sept (Ratmanov 1937a, b)	*Dal'nevostochnik*	Gulf of Anadyr, around St. Lawrence to Pt. Hope	Hydro stations, currents in Bering Strait
1933	24 Jul–28 Aug (Ratmanov 1937a, b)	*Krasnoarmeyetz*	Gulf of Anadyr, south and north of St. Lawrence north to Herald Island	Hydro stations, currents in Bering Strait
1934	30 Jul–6 Aug (Barnes and Thompson 1938)	*Chelan*	North of St. Lawrence through Bering Strait	Hydro stations, currents in Bering, Anadyr Straits
1937	13 June–19 Sept (Goodman et al. 1942)	*Northland*	Norton Sd., St. Lawrence Is. north to Wainwright	Hydro stations
1938	3 Aug–5 Sept (Goodman et al. 1942)	*Northland*	St. Lawrence Is. north to Pt. Barrow	Hydro stations
1949	Jul–Aug (Saur et al. 1954–water masses) (Lesser and Pickard 1950–currents)	*Cedarwood*	St. Lawrence Is. to 73°N east of 1867 Convention Line	Hydro stations, currents in Bering Strait
1951	1–22 Feb (U.S.N.H.O. 1954)	*Burton Island*	East, west, and north of St. Lawrence	22 hydro stations
1955	15 Mar–4 Apr (U.S.N.H.O. 1958)	*Northwind*	SE, E, NE of St. Lawrence	13 hydro stations
	3–26 May (U.S.N.H.O. 1958)	*Burton Island*	SE of St. Lawrence to King Is.	7 hydro stations
1959	2 Aug–1 Sept (Fleming and Heggarty 1966)	*Brown Bear*	Bering Strait to Cape Lisburne	178 hydro stations, current measurements
1960	26 Jul–7 Aug (Fleming and Heggarty 1966)	*Brown Bear*	St. Lawrence to Pt. Hope (east of Convention Line)	Hydro survey; 5, 20 m currents
1962	5–26 Oct	*Northwind*	Chukchi Sea	110 hydro stations

Year	Dates (Reference)	Ship	Location	Description
			Anadyr and south of St. Lawrence	no. 1, 1964)
1963	7–12 Aug (U.S.C.G. Oceano. Rpt. no. 6, 1965)	Northwind	Bering Strait to Wrangel Is.	41 hydro stations
1964	5–7 Aug (Coachman and Aagaard 1966)	Northwind	Bering Strait	Section of anchored stations, 3 day current mooring
1966	30 Jul–24 Aug (Coachman and Rankin 1968)	Burton Island	Long Strait	Current measurements
1967	12–17 Jul (Husby 1969)	Northwind	South of Bering Strait to Pt. Hope	4 sections of anchored stations: T, S, currents
	18–23 Jul (Husby 1969)	Northwind	(Repeated occupation of above survey)	
	2–7 Aug (unpublished)	Thompson	Bering Strait	4 repeated sections anchored: T, S, currents
1968	2–19 Feb (Countryman and Bourkland 1968)	Northwind	Gulf of Anadyr Strait of Anadyr	T, S, currents
	9–19 Jul (Husby 1971)	Staten Island	St. Lawrence Is. to north of Bering Strait	6 sections of anchored stations: T, S, currents
1969	14 Feb–3 Mar 5–25 Apr (unpublished)	Staten Island	East, south, west of St. Lawrence, north through Bering Strait	T, S and currents
	8–21 Jun (Husby and Hufford 1971)	Northwind	St. Lawrence to north of Bering Strait	T, S sections (same as summer 1968)
1970	2–16 Aug (Hufford and Husby 1972)	Northwind	Gulf of Anadyr	T, S, 5 current stations
	23 Sept–18 Oct (Ingham et al. 1972)	Glacier	NW, N, NE of C. Lisburne	92 hydro stations, with 12 current stations
1972	Jul–Aug (unpublished)	Oshoro Maru	Bering Strait north to Wrangel Is.–Pt. Barrow	4 sections of anchored stations: T, S, currents
1973	29 Sept–4 Oct (unpublished)	Thompson	Bering Strait	2 sections of anchored stations in Bering Strait, 1 south of strait: T, S, currents

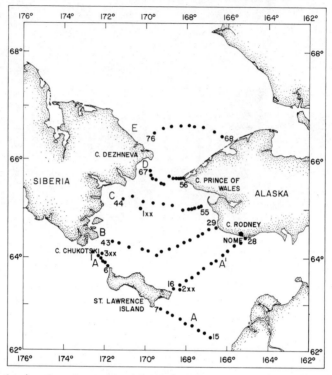

3. Station locations *Staten Island* 9–19 July 1968

and current stations in the Strait of Anadyr and westward into the Gulf; and the 1969 *Staten Island* cruise occupied 67 stations west, south and east of St. Lawrence, as well as northward on the eastern side and on through Bering Strait. Four days of current records were obtained in March 1968 from the ice west of Kotzebue Sound (Coachman and Tripp 1970).

2

WATER MASSES

WE BEGIN BY DEFINING and describing the water masses found in Bering Strait. Temperature-salinity diagrams for certain of the detailed (12 to 13 stations) cross-sections are shown as Figs. 4 to 7. In these sections, all taken in midsummer, station spacing was 5 to 8 km. and the normal vertical sampling interval 5 m. The section location (cf. Figs. 2, 3) is: west from Cape Prince of Wales, past Fairway Rock, to south of Little Diomede Island, then (on a line running from 22 km south of Ratmanov Island) NW to 22 km from the Siberian shore, SSW of Cape Peek. While this section configuration has been dictated by the contemporary international political situation, we believe that only a small fraction of the flow is missed.

The most striking feature of the T-S relationships for each station is their orientation largely parallel to the T-axis; i.e., variations in any column of water are largely in temperature. The deepest observation (in practically every case) was the coldest, the water column becoming warmer toward the surface but with little variation in salinity. In some few cases, e.g., stations 2 to 7 in Fig. 4, the T-S curve does show a slope upward toward the left, evidencing increased vertical salinity differences, as the near-surface layers are both warmer and less saline than those below. However, this is not the typical situation except at stations close to Cape Prince of Wales. Thus, in midsummer vertical differences large in temperature but small in salinity is the general rule. Maximum surface temperatures may be higher some years than others (in 1964: 10.5°C; in 1968: 8°C), but the vertical range in any year does not exceed about 8°C.

Thus, the significant variation in salinity is laterally across the strait, with the most saline water always to the west and the salinity diminishing progressively toward the east. The range of salinity encountered is in general $<3°/_{00}$, but the minimum and maximum salinity values differ somewhat from year to year. Thus, the minimum in 1964 was $\sim 30°/_{00}$ but in 1968 $\sim 31.5°/_{00}$. The maximum varies over narrower limits, from just less than $33°/_{00}$ (1972) to $33.25°/_{00}$ (1967) in these data.

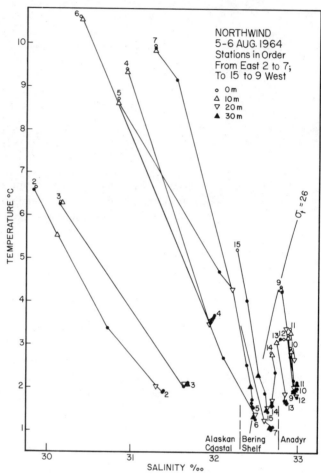

4. Temperature-salinity diagram for stations of Bering Strait section, *Northwind*, 5–6 August 1964: station numbers in order from east to west are 2 to 7, 15 to 9

PRINCIPAL WATER MASSES

A marked feature of the T-S distributions of midsummer is the grouping into bands of relatively narrow salinity ranges (station spacing was quite uniform; cf. Figs. 2a, b, 3). These bands are generally discrete, overlapping only occasionally near the surface. The salinities of the deepest observations thus concentrate into three groups. As an example, some of the Bering Strait stations have been plotted in Fig. 8, with envelopes drawn around all observations from groups of adjacent stations.

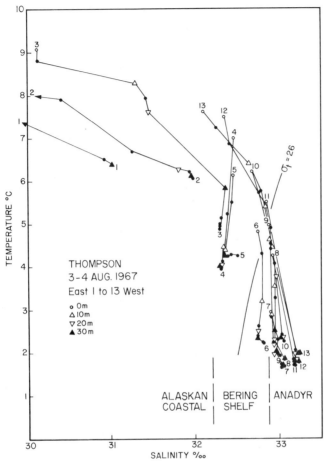

5. Temperature-salinity diagram for stations of Bering Strait section, *Thompson*, 3–4 August 1967: station numbers in order from east to west are 1 to 13

The figure clearly shows the three discrete lobes projecting toward cold temperatures.

Thus, based on salinity, we define three water masses in Bering Strait and assign them the names *Anadyr*, *Bering Shelf*, and *Alaskan Coastal*, which are descriptive of their sources. Contrary to the usual practice, the water masses are not prescribed by fixed limiting values of salinity, but rather by floating ranges because of the obvious year-to-year differences in values. These ranges are indicated in Figs. 4 to 7. With emphasis on the deeper salinities, approximate median values can be selected for each water mass (Table 2).

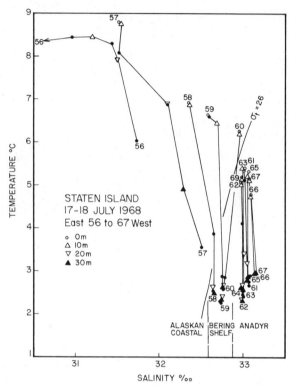

6. Temperature-salinity diagram for stations of Bering Strait section, *Staten Island*, 17–18 July 1968: station numbers in order from east to west are 56 to 67

TABLE 2
Approximate Median Salinities of Water Masses, ‰

Summer Section in:	Alaskan Coastal	Bering Shelf	Anadyr
1964	31.7	32.6	32.95
1967 (3–4 Aug)	31.5	32.5	33.1
1968	32.1	32.7	33.1
1972	31.9	32.4	32.85

The only previous water mass classification scheme proposed for this region has been that of Saur, Tully, and LaFond (1954). Their classification was based on the summer 1949 data obtained from the *Cedarwood*, which occupied some 164 hydrographic stations from just south of St. Lawrence Island northward through Bering Strait to 73°N, but staying east of the Convention Line of 1867 (Fig. 1). Thus, their classifica-

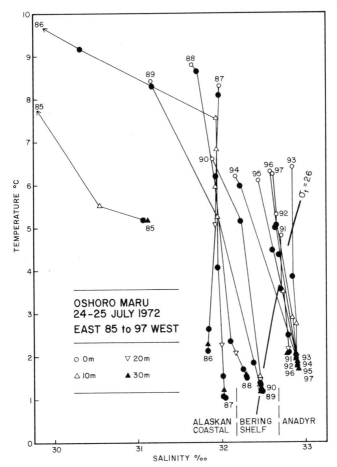

7. Temperature-salinity diagram for stations of Bering Strait section, *Oshoro Maru*, 24–25 July 1972: station numbers in order from east to west are 85 to 97

tions were based not only on data from Bering Strait and the northern Bering Sea, but also from the Chukchi Sea. While water from the northern Bering Sea is found in Bering Strait (see below), so that observations from south of the strait would properly contribute to a useful water mass classification scheme for Bering Strait, data from the Chukchi Sea by and large do not so contribute, because only on rare occasion is water from the north found in the strait.

Figure 9 shows those T-S classifications of Saur et al. which apply to Bering Strait. The classification "Deep Shelf Water" would be appropriate in winter, but clearly not in summer; it probably is based on deep

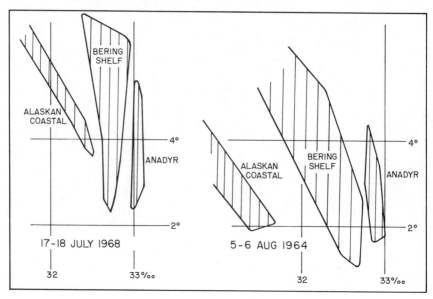

8. Water mass envelopes for the Bering Strait section for 17–18 July 1968 (left, cf. Fig. 6) and 5–6 August 1964 (right, cf. Fig. 4)

observations from the Chukchi. Three other classifications apparently also based on Chukchi Sea water have been omitted. For comparison of our water mass description with this previous scheme, the detailed 1972 *Oshoro Maru* data (Fig. 7) are superimposed. The *Cedarwood* data are not available, and were not taken in the same detail near Bering Strait to give coverage comparable to that of the recent sections. Thus, we have only the envelopes for comparison. The 1972 data were selected because the absolute values were similar to those obtained in 1949.

In our view, the T-S classification scheme of Saur et al. does not group the observations in a manner which provides insight into the physical behavior of the system. Their scheme is based on temperature differentiation, whereas the data demonstrate distinct differentiation primarily in salinity, suggesting approximate conservation of salt within each water mass. This is particularly true with the deep salinities, and frequently the separation extends to the surface. On the other hand, heat is not conserved, and variable amounts are entering and leaving the water columns.

Thus, we reach a basic conclusion about the behavior of water masses in the region, viz., there has been as yet very little lateral mixing. The water masses apparently attain their characteristic salinities upstream (to the south, because of the predominant northward flow) and subsequently maintain their characteristic salinity separation, as is evident from adjacent stations <5 km apart. Occasionally stations will be taken in water with transitional characteristics; e.g., 1964 stations 13 and 14 (Fig. 4) and 1967 station 6 (Fig. 5) demonstrate the lateral transition between Anadyr and Bering Shelf water masses, and 1968 station 57 (Fig. 6) shows the transition between Bering Shelf and Alaskan Coastal waters.

There is at certain times and under certain conditions (e.g., under the influence of wind) some lateral mixing near the surface, leading to vertical salinity differences in addition to the normal summer temperature differences. For example, in 1964 (Fig. 4) within the Bering Shelf and Alaskan Coastal water masses some such mixing had occurred, while this effect was less noticeable in 1972 (Fig. 7) and hardly at all in 1968 (Fig. 6).

WATER MASS SOURCES

Suitable data for examining the water mass structure and distribution upstream (south) from Bering Strait are available from 1968 *Staten Island* (Fig. 3) and 1969 *Northwind* (1968 stations duplicated). In both cruises, sections were occupied from south to north, tending to minimize advective field changes during the cruise, and the region was covered in less than two weeks.

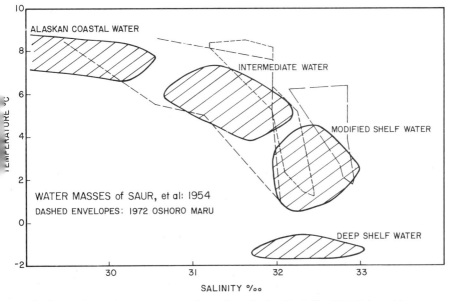

9. Comparison of water mass envelopes from Bering Strait (for 1972 *Oshoro Maru*, cf. Fig. 7) with water masses of Saur, Tully, and LaFond (1954)

Figure 10 presents T-S plots of the 1968 stations across the Strait of Anadyr (stations 1 to 6, section A in Fig. 3) and from the eastern end of St. Lawrence Island ESE to the Alaskan coast (stations 7 to 15). The T-S curves show the same patterns as those from Bering Strait, forming three separate groupings on the basis of the deeper salinities. In the Strait of Anadyr, stations 1 to 4 show the highest salinities (32.9 to 33.0‰) while stations 5 and 6, which lie closest to St. Lawrence Island, fall in the range 32.5 to 32.8. East of St. Lawrence, stations 7 to 10 form a group with salinities identical with 5 to 6 (32.5 to 32.75); and stations 11 to 15 had the lowest values.

We identify these as the same three water masses distinguished in the

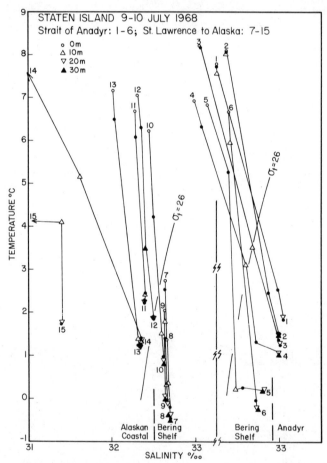

10. Temperature-salinity diagram for stations from a closed section from Siberia to Alaska encompassing St. Lawrence Island: station numbers in order from east to west are 15 to 7 (St. Lawrence Is.), 6 to 1 (cf. section A in Fig. 3)

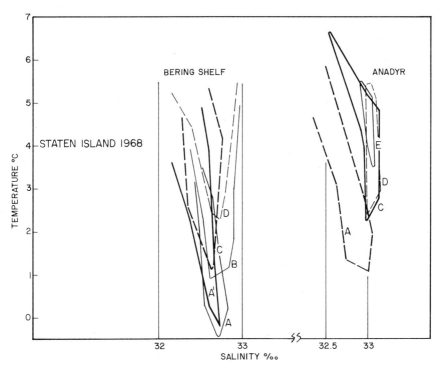

11. T-S envelopes of Bering Shelf (left) and Anadyr (right) water masses from 1968 *Staten Island* sections lettered sequentially from south to north (cf. Fig. 3)

Bering Strait data. To demonstrate the continuity in the system from St. Lawrence Island to north of Bering Strait, Fig. 11 shows the T-S envelopes enclosing the Anadyr and Bering Shelf water masses section by section (sections lettered in Fig. 3). As both water masses are advected northward through the system, their minimum temperatures steadily increase. Thus, the minimum temperature of Anadyr Water in section A was ~1°C, in section C ~2.2°C, in D ~2.4°C, and in E, north of Bering Strait, ~3.5°C.* Likewise, the minimum temperature of the Bering Shelf Water increased regularly from −0.5°C by St. Lawrence Island to 2.3°C in Bering Strait. The salinity bands, on the other hand, remained very much the same. We cannot expect precise coincidence of the salinities because there are actually only a few stations in each water mass in each section, even in this detailed survey. For example, the section A Anadyr Water (Fig. 11) is not quite as saline as later sections (C, D), quite likely because the water with higher salinity lay west of the end of the section.

*An envelope for section B is not shown because only the westernmost station (43) was entirely in the Anadyr Water, the bulk of which probably lay to the west of station 43 (which was ~31 km from the coast), while the next four stations to the east (39 to 42) were mixed Anadyr-Bering Shelf Water.

The spatial distribution of these water masses from St. Lawrence to north of Bering Strait is pictured in Fig. 12. There is a continuity of each water mass within the system, with Anadyr Water on the west, Alaskan Coastal Water on the east, and Bering Shelf Water between. In certain regions there is significant mixing between water masses, notably northeast of St. Lawrence in section A (two stations transitional between Bering Shelf and Alaskan Coastal Water); and north of western St. Lawrence in section B (five stations with mixed Bering Shelf-Anadyr characteristics). However, the major feature of the distributions is the distinct separation maintained through the region, which confirms the conclusion that in general lateral mixing is small.

Included in Fig. 12 are lines connecting stations with the minimum temperature in each water mass. These can be considered the paths of the core of each water mass and will be used in later calculations.

Two of the water masses, Anadyr and Bering Shelf, originate south of St. Lawrence Island. Alaskan Coastal Water also has a contribution from the south, but there is also substantial input from Norton Sound.

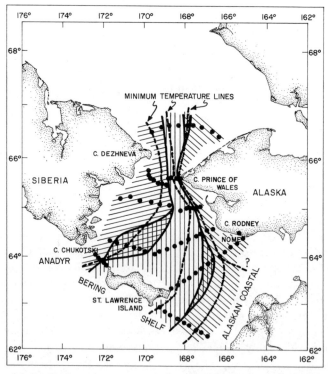

12. Spatial distribution of water masses for 1968 *Staten Island* (Dashed lines connect stations which showed minimum temperature in each water mass.)

Anadyr Water

At the Strait of Anadyr, there is an intimate connection between Anadyr Water and the Gulf of Anadyr to the southwest (Fig. 12). Ohtani (1969) has described the characteristics of the water overlying the whole Bering Sea shelf, based on July, 1964 to 1966, data from *Oshoro Maru*, and his distributions of temperature and salinity at the bottom are presented in Fig. 13. As the water column in summer typically exhibits two layers separated by a sharp pycnocline, and the bottom layer is nearly homogeneous, the distributions shown in Fig. 13 are representative of a water layer several 10s of meters thick.

The isohalines suggest a continuity of water $>33°/_{00}$ from the deep Bering Sea basin, south of Cape Navarin, then along the Siberian shore curving eastward toward the Strait of Anadyr. This water is warmer ($>1°C$) than the waters bordering to the southeast or northwest ($<0°C$). However, the stations are few and widely spaced, and more detailed data are required to be certain of the water mass distributions.

Three relatively detailed surveys are available from the Gulf of Anadyr: *Northwind* 1962 (Gladfelter et al. 1964); *Northwind* 1968 (unpublished); and *Northwind* 1970 (Hufford and Husby 1970). The 1962 data were taken in September and the 1968 data in February, while those from 1970 were acquired during midsummer (July). We shall initially use the latter for discussing the Anadyr Water, as they provide the best comparison with the Bering Strait conditions described so far, due to seasonal variations in properties (see below).

The 1970 stations are shown in Fig. 14, and a T-S diagram of certain station groupings in Fig. 15. One grouping (stations 8 to 10, 20 to 22, 25, 26) was composed from all stations with deep water characteristics analogous to those of the Anadyr Water, viz., temperatures above 1°C and salinities close to $33°/_{00}$. The dashed lines in Fig. 14 include these stations.

The only station from the Strait of Anadyr, 34, obviously was not in the same water mass. It was located about 55 km from the Siberian coast and was therefore too far east to show the Anadyr Water; rather, it represents Bering Shelf Water (cf. Fig. 10).

As a hypothesis, we now suppose that the Anadyr Water is not endemic to the Gulf of Anadyr, but originates in the northern Bering Sea and transits the gulf on its northward journey. Its relative warmth (1 to 2°C) and high salinity ($>33°/_{00}$) in the deep layer, compared with the other Bering Strait region water masses, are due to its origin in more southerly latitudes. In support of this concept, Fig. 16 shows an expanded-scale T-S plot of selected stations. The station exhibiting the warmest and most saline deep Anadyr Water is station 8, from the southwest corner of the

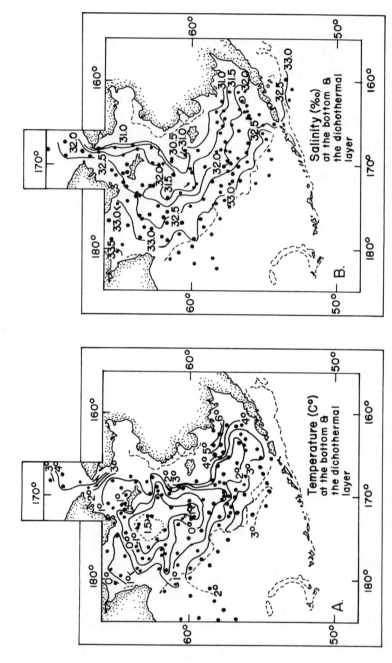

13. Horizontal distributions of temperature °C (A) and salinity ‰ (B) in the near-bottom layer, based on *Oshoro Maru* data from midsummers of 1964–66 (from Ohtani 1969)

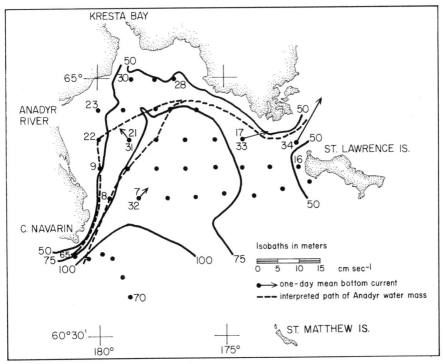

14. Station locations in Gulf of Anadyr for 1970 *Northwind*, stations 1 to 34, and 1972 *Thompson*, stations 65 to 70. (Arrows are vectors of measured bottom currents, and dashed lines interpreted path of flow of Anadyr water mass.)

grid. Following the core of the Anadyr Water along these stations (cf. Fig. 14), 10, 20, and 26, shows the deep layer to be progressively colder and less saline. The source of water for mixing to cause this change is to the east, in the central Gulf of Anadyr, represented by stations 6, 17 to 19. Stations 7, 11, and 27 along the east side of the core of the Anadyr Water, and west of the cold central gulf, show transitional character, with a layer of relatively low temperature and salinity overlying Anadyr Water. At these transitional stations, there has also been some deep mixing, as the bottom values are somewhat colder and less saline than the usual Anadyr Water.

The end result of the mixing, i.e., the composition of Anadyr Water at the Strait of Anadyr, can be estimated as 80 to 90% water from the Bering Sea and 20 to 10% water from the central Gulf of Anadyr. On the other hand, the water lying north and west of the core (stations 23, 24, 29, 30) is quite cold and saline below 30 m and appears to have very little interaction with the transient Anadyr Water. Also shown in Fig. 16 are the 25 and

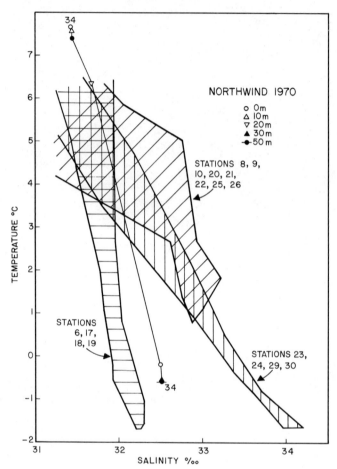

15. T-S envelopes for groups of stations from Gulf of Anadyr, 1970 *Northwind*

30 m observations from stations 24, 29, and 30 showing that at these intermediate depths there was a layer of Anadyr Water above the much colder and more saline deep water of the northwest part of the gulf. Little interaction between these water masses, either lateral or vertical, can be seen from these data, although the 1970 survey was not spaced closely enough to detect the more subtle features of the water mass interactions.

The Bering Sea water intrusion, which gives rise to the Anadyr Water, enters the Gulf of Anadyr close to Cape Navarin. The 1970 *Northwind* data did not extend south to Cape Navarin, but a section of six stations was made in summer 1972 from the *Thompson* from 22 km south of Cape Navarin toward the southeast (stations 65 to 70 in Fig. 14). Figure 17

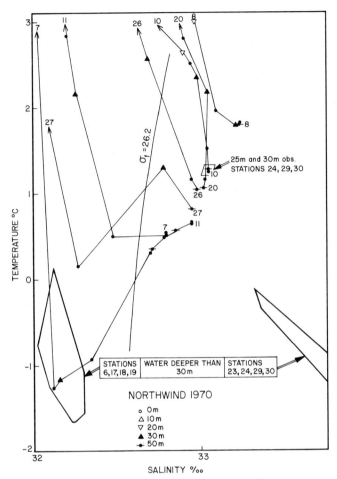

16. T-S diagram of stations from Gulf of Anadyr, 1970 *Northwind*, and envelopes for observations deeper than 30 m (two groups) and for 25 and 30 m observations (one group). (Note expanded scales.)

shows T-S plots from this section. Though the values may not coincide exactly with those from 1970, the westernmost stations (65 and 66) are the only ones showing properties appropriate to the primary parent water mass of the Anadyr Water; i.e., the water column from 50 m down had a nearly constant salinity of ~33.2‰ and temperatures between 1.75° and 2°C. The next station east (67) showed a temperature minimum and influence of water from the central gulf. The most extreme influence was at stations 68 and 69 (32.2 to 32.3‰ and −1 to −1.5°), while station 70 to

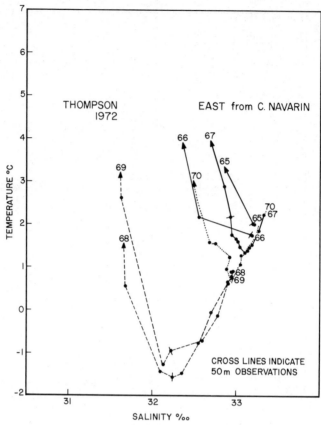

17. T-S diagram of stations 65 to 70, 1972 *Thompson* east from Cape Navarin

the southeast showed less influence, but still cannot be confused with the progenitor of the Anadyr Water.

The Anadyr Water lies somewhat deeper near Cape Navarin than it does within the gulf: near Cape Navarin it occupies the water column below 50 m (cf. Fig. 17), whereas within the gulf it extends about 20 m higher (cf. Fig. 16).

The 1970 *Northwind* cruise also obtained five bottom current measurements, which agree with the flow regime we have inferred from the T-S distributions. At stations 7, 17, 21, 30, and 34 near-bottom currents were measured from the anchored vessel at frequent intervals for station durations of more than 25 hours (Husby 1973). The 24-hour mean flows at four of the stations are shown in Fig. 14. [The current at station 30 was strong ($\sim$15 cm sec^{-1}), but oscillated between north and south, with an average near zero. There is a large ($\sim$6 m) tide range in Kresta Bay

(Lisitsyn 1966), and though the current record is too short to be definitive, the N-S orientation of a strong oscillating current at the mouth of this bay indicates a primarily tidal character to the flow.] There was some variability in the record from station 21, while stations 7, 17, and 34 showed little variability in speed and direction. These direct current measurements agree with the flow field inferred from the hydrography.

We conclude that the Anadyr Water of the Bering Strait region originates in the Bering Sea, probably as a subsurface partial continuation of the flow which parallels the continental slope of the deep central Bering Sea basin from southeast to northwest (the Bering Slope Current, cf. Kinder et al. 1975). Unlike some previous depictions of this flow (Lisitsyn 1966; Arsen'yev 1967), the water does not enter the Gulf of Anadyr from the south on a broad front in summer; rather its core appears to lie close to Cape Navarin.

In transiting the Gulf of Anadyr, the water follows the isobaths. In summer the water mass retains to considerable degree its relatively warm temperature (>1°) and high salinity (>32.9°/$_{00}$). It is slightly modified by mixing on the south and east sides with a very cold (<−0.5°C) and less saline (<32.3°/$_{00}$) water mass occupying the central gulf. The net effect is a cooling (of up to 1°C) and a slight freshening (0.1 to 0.2°/$_{00}$) during transit, resulting in the Anadyr Water of the Bering Strait region, which is perhaps 80 to 90% Bering Sea water and 20 to 10% central gulf water. A similar analysis of the 1962 *Northwind* data confirms these conclusions in every respect.

A most intriguing aspect of the Gulf of Anadyr regime is the selective mixing between water masses. The Anadyr Water mixes relatively little (preferentially along the south and east sides of the flow rather than along the north and west sides). Thus, the turbulent flow regime on the two sides of the flow must differ. We note that the mixing takes place largely along isopycnal surfaces, as shown by the trend of the deep core characteristics at stations 8, 10, 20, and 26 (Fig. 16) running nearly parallel to the σ_t lines. This may be the reason why little mixing occurs with the water to the north and west, which is distinctly denser. The situation will be contrasted with the mixing regime of the central Gulf of Anadyr in the next section.

Bering Shelf Water

This water mass, with characteristic temperatures between −0.5 and 1 to 2°C and salinities of 32.5 to 32.8°/$_{00}$, was seen to enter the Bering Strait region from around both ends of St. Lawrence Island (Fig. 10). It is not the same water as that found in the central Gulf of Anadyr (cf. Fig. 15) which is both colder and less saline. However, the intermediate position of station 34 on the T-S diagram (Fig. 15) suggests, as an initial hypothesis, that the Bering Shelf Water is also a mixture of water from the Bering Sea with the less saline, very cold water resident on the northern Bering Sea shelf. The creation of this water mass by mixing of Bering Sea water would be different in kind and degree from the relatively minor modification in transit of Anadyr Water.

Examination of Fig. 13 shows the very cold water to have occupied a large but isolated position southwest of St. Lawrence Island in 1964 to 1966. The salinity is between 31.5 and 32.5$°/_{00}$ (Fig. 13B). It appears that this water is similar to the central Anadyr water described by the 1970 *Northwind* data, though at the latter time the cold center had penetrated farther to the northwest. Figure 18 shows the distribution of the cold water at yet another time, September 1962, when its position coincided closely with that defined by the *Oshoro Maru* data. The feature is, in fact, the "cold center" of the Bering Sea shelf already described by Barnes and Thompson (1938).

The formation of a cold resident water occurs in winter due to cooling

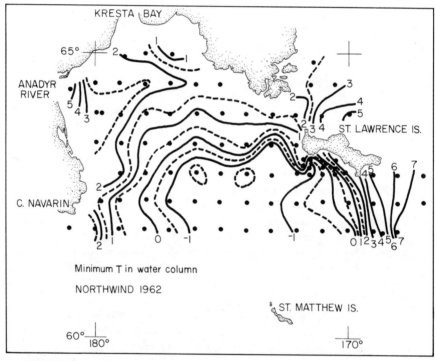

18. Minimum temperature °C observed in water column, 7–18 September 1962, *Northwind*

and formation of ice. The upper layer of warm and low-salinity summer water is destroyed during fall both by convection and by flushing. On the surface, where salinities are ~32‰, 1 to 2 m of ice is formed during winter, which can raise the salinity to 33‰ or more.* The water would be thoroughly mixed, top to bottom, and the O_2 content initially near saturation (see also *Seasonal Changes* below). This cold water is subsequently isolated from the surface by the development in early summer of the less dense upper layer. In those areas without a vigorous horizontal circulation, this winter-formed resident water readily retains its identifying characteristics over the summer.

As the 1962 *Northwind* data are the most comprehensive available from the Gulf of Anadyr, we examine them more closely. The stations were grouped into (a) those with $S_{max}>33$‰ and no temperatures colder than 1.75°C and (b) those in which there were no salinities >32.5‰ and with

*Growth of 2 m of ice in a water column of 32‰ 50 m deep would increase the salinity to 33.1‰.

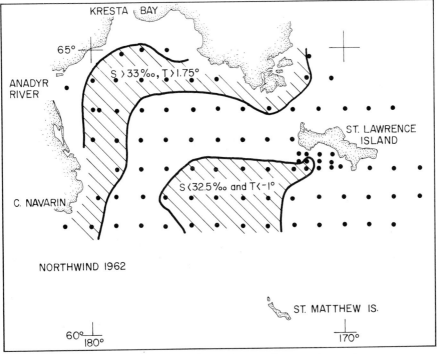

19. Station groupings in Gulf of Anadyr of those with $S_{max}>33$‰ and no temperatures <1.75°C, and those with no salinities >32.5‰ and $t_{min}<-1°C$, 1962 *Northwind*

$t_{min} < -1°C$. The distribution of the groups is shown in Fig. 19 and T-S envelopes of the groups in Fig. 20. The first group has the characteristics of Anadyr Water, and its distribution within the gulf was similar to that previously described for this water mass. The second group constitutes a very cold, less saline central Anadyr Gulf water mass, and its distribution is that of the cold spot.

Also included in Fig. 20 are stations 7 and 61. These identify the Bering Shelf Water, which had only slightly lower temperatures than those of the Anadyr Water, but showed a characteristic salinity $\sim 0.3‰$ lower.

All stations lying between groups (a) and (b) (cf. Fig. 19) were examined, and representative ones are plotted in an expanded-scale T-S

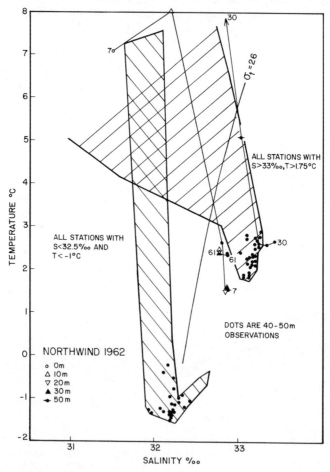

20. T-S envelopes of station groups of Fig. 19

diagram in Fig. 21. The characteristic feature of all these intermediate stations is the hook shape of the deeper layers; i.e., the water column has an intermediate temperature minimum at a salinity somewhat lower than that of the bottom water. Such a T-S pattern must be predominantly due to and maintained by both layering and lateral mixing.

Consider two adjacent water columns, both with a similar warm and low-salinity uppermost layer. At station A the deeper water is of the same salinity and very cold, while at B the deeper water is more saline and warmer. The relationships are sketched on the T-S plane in Fig. 22. Extremes of possible interaction between A and B would be (1) a 50-50

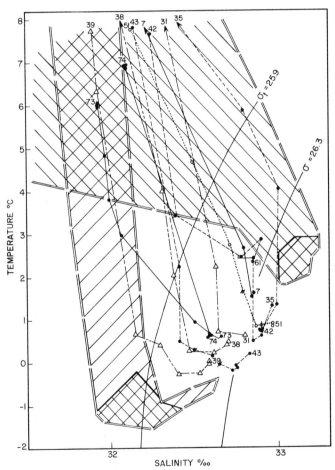

21. T-S envelopes of station groups of Fig. 19 with expanded salinity scale and T-S diagram of stations located between the groups

32 • BERING STRAIT: The Regional Physical Oceanography

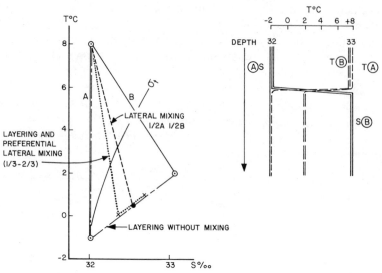

22. Schematic diagram of layering and lateral mixing

mixture of the waters at all depths and (2) an intrusive layering, with density increasing down but without mixing. The T-S curves for these cases are shown, and neither exhibits the characteristics of the stations of Fig. 21.

Now consider a case where there is both intrusive layering and mixing. With layering, the deepest layer would be dominated by the warm saline water while the intermediate level would be dominated by the cold and less saline water. Mixing would then modify each layer according to the dominant water mass. Thus, one water column could have a bottom layer of 1/3A, 2/3B, so that the resulting mixture in the bottom layer would be relatively warm and saline (and densest), while that in the layer above would be 2/3A, 1/3B, and therefore relatively cold and less saline (and less dense). The resulting T-S curve (Fig. 22) conforms closely to those observed. The amount of layering and mixing would vary downstream, i.e., the resulting water columns would exhibit a progressively more mixed character, and, therefore, a diminishing hook pattern on the T-S plane. Likewise, these processes would vary cross-stream, and the T-S curves would reflect the dominance of one or the other parent water mass, depending on the station's location relative to them.

In Fig. 21, such cross-stream variation is shown for stations located on lines between the cold spot and the Anadyr Water, viz., 39, 38, 31 and 43, 42, 35. The downstream variation is indicated by progressive changes from the former group to the latter and on to station 51. Therefore the flow direction must be from the southwest part of the Gulf toward the north and east, parallel to the flow of the Anadyr Water. The densest source is the Anadyr Water, which always lies at the bottom. Modifications of this water are small, because there is no water of an equivalent density with which to mix. In the layer above, there are more nearly equal volumes of water in the same density range from both parent water masses, so that the degree of interaction is markedly greater. Ultimately, this leads to the disappearance of the hook in the T-S curve (cf. the transition from stations 39 to 51 to 7).

By the time the waters have completed their transit of the gulf, the water columns are relatively isohaline, but are in a salinity band distinguishable from that of Anadyr Water (i.e., station 7). We note that past the Strait of Anadyr, the mixing character of the regime apparently changes, with lateral mixing becoming much less important. The change from stations 7 to 61 shows the effects of vertical mixing, but not lateral; and, as noted earlier, north of St. Lawrence Island lateral mixing seems to be unimportant. This situation seems to be correlated with the slope of isopycnal surfaces in the region. South of the Strait of Anadyr, the average slope is relatively small, typically being $\sim 2 \times 10^{-4}$ in the region of the Bering Shelf Water. In the strait and to the north, the slope increases by at least two orders of magnitude, to $\sim 5 \times 10^{-2}$. Where the isopycnal surfaces are close to horizontal, lateral mixing assumes greater importance, vis-à-vis vertical mixing, than where the surfaces are greatly inclined to the horizontal.

A quantitative estimate of the lateral mixing can be made from these data by the method of Proudman (1953, p. 126). For stationary lateral mixing

$$\frac{Ky}{U} = \frac{1}{8} \frac{(\delta y)^2}{\delta x},$$

where Ky is the lateral eddy coefficient, U the mean flow, δx the distance separating the sections in line of the flow, and δy the distance across the core in the T-S plane as defined by Proudman (p. 126). The two cross-stream sections of stations 34, 35, 42 and 17, 13, 42 appear to give a reasonable depiction of the cross-stream property distribution at 50 m depth, from which Ky/U was estimated as 0.5×10^6 cm. If U was 5 cm sec^{-1}, Ky was 2.5×10^6 cm^2 sec^{-1}.

It has been shown (Okubo and Ozmidov 1970) that lateral eddy coefficients increase with increasing size scale of the system. The lateral scale in the present problem is of order 100 km, for which their results give $Ky = 1.2 \times 10^6$ cm^2 sec^{-1}. This is in close agreement, if U is of order 1 to 10 cm sec^{-1}.

Thus the evidence all suggests a general clockwise circulation following the isobaths around the cold spot. The Anadyr Water remains close to the Siberian shore, and in summer passes *in toto* through the Strait of Anadyr. The Bering Shelf Water apparently bifurcates near the Strait of Anadyr, part passing through the strait and part entering the region south of St. Lawrence Island. For example, Fig. 21, stations 73 and 74 appear to be downstream from station 39.

An aspect of the conditions observed in September 1962 which appears to differ from midsummer is that no Bering Shelf Water was found near the eastern end of St. Lawrence, whereas in summer this water enters the Bering Strait region from around both ends of St. Lawrence. This suggests that earlier in the season there may have been a much larger pool of Bering Shelf Water south of St. Lawrence, part of which probably comes from the Gulf of Anadyr in the manner described above. However, as there cannot be advection across a region of a conservative property minimum such as the cold spot, the circulation is around it. Mixing to create the Bering Shelf Water may thus also occur throughout much of the region immediately south of St. Lawrence Island.

These results also illuminate the origin of the cold spot. The general distribution of salinity in the deeper water on the shelf (Fig. 13b) shows a tongue of low-salinity water entering the region between St. Lawrence and St. Matthew from the east, passing close to St. Matthew, on the south side of the anticyclonic flow. If this water is mixed with the warmer but more saline water from the Bering Sea, the resulting mixtures would be less saline and therefore less dense than Bering Sea water proper, but denser than any other water resident on the shelf. Ice formation in winter and the attendant vertical convection generated thereby, and extending to the bottom in these shallow depths (as discussed by Ohtani 1969), would tend to maintain the relatively high salinities but would remove heat from the water column, giving rise to the cold temperatures. In summer the bottom layer becomes insulated against rapid heat loss by the development of the near-surface pycnocline, and, in the absence of advection, remains isolated as a pool of colder water.

Figure 23 shows the O_2 percent saturation of the deeper water, based on the 1962 *Northwind* data. The generally low values (~70%) coinciding with the cold spot lend credence to the above hypothesis. The low saturation values in the northern and northwestern Gulf of Anadyr also illustrate the relative lack of interaction of the deeper water there with the Anadyr Water, as deduced previously. The oxygen utilization rate in the cold spot can be estimated, providing the formation and seasonal history hypotheses are correct. Winter vertical convection (see below) not only will maintain the temperature near freezing, but would also

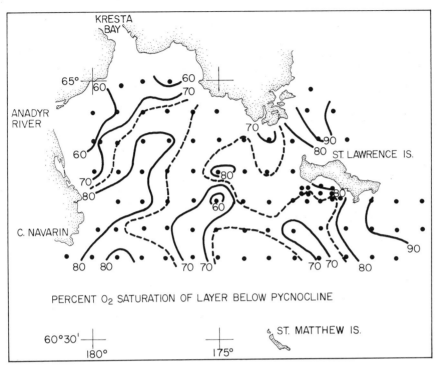

23. Percent saturation of oxygen of the layer beneath the pycnocline, Gulf of Anadyr (1962 *Northwind*)

aerate the water. This convection ceases about the beginning of April, at which time the water should be approximately saturated with O_2, about 8.4 ml/l. In September 1962, five and a half months later, values were about 5 ml/l, which suggests a net utilization rate of 7 to 8 ml/l/yr.

In conclusion, we have identified the sources of two major water masses of the Bering Strait region. They are the Gulf of Anadyr and the region south of St. Lawrence, where waters from the Bering Sea are variously mixed and modified. These modifications and the general features of the flow regime have been described, but the data available are insufficient to elucidate all details of the flow and mixing. For example, the detailed nature of the flow field near St. Lawrence Island is probably complex: there is evidence from sediment distribution (Knebel 1972) that the mean flow quite close to the south side of St. Lawrence Island is to the west. More importantly, the oceanography of the interesting (particularly to the fisheries) region between St. Lawrence and St. Matthew demands much more study than has been undertaken to date.

Alaskan Coastal Water

This water mass is the warmest and least saline water in the system. Its T-S curves are directed from relatively cold and saline water near the bottom, toward warmer and fresher water at the surface, without marked bends or kinks in the curves. Thus, we interpret this water mass in summer as formed from cold water of approximately $32°/_{00}$ salinity, which has subsequently been diluted to varying degree by fresh water from the Alaskan coast and warmed largely by solar radiation.

The 1960 *Brown Bear* cruise provided a quasi-synoptic survey of the whole eastern part of the region, and the 5 m temperature and salinity distributions, which well represent the surface layer values, are shown in Figs. 24a, b (Fleming and Heggarty 1966). The salinity distribution clearly demonstrates the direct connection between Alaskan Coastal Water and the major source of fresh water in the region, the Yukon River. The low salinity water is advected north along the coast, and the salinity progressively increases as salt is diffused upward from the more saline deeper layer. A similar pattern occurs in the vicinity of Kotzebue Sound which is supplied by the Kobuk and Noatak rivers, whence the low salinity water is advected north along the coast past Pt. Hope.

The same flow features can be interpreted from the 1968 *Staten Island* data, though the salinity values were higher ($29°/_{00}$ near the Yukon), and the $31°/_{00}$ isohaline lay closer to the Alaska coast than in 1960. The range of values achieved during any season depends directly on the amount of runoff available to mix with the saline parent water. This amount varies year to year, and in 1960 the runoff influencing the region at the time of the survey had been about average, while in 1968 it had been well below average (see *Year-to-Year and Long-Term Variations*).

The deeper water within the Alaskan Coastal Water also exhibits a range of salinity values. Figure 13b illustrates the near-bottom layer salinity gradient across the shelf, with the salinity progressively diminishing as the Alaskan coast is approached. Rather thorough vertical mixing will take place in winter, when, according to Ohtani (1969), vertical convection reaches to the bottom, or to deeper than 30 m in the more northerly areas. No thorough survey of the Bering Sea shelf in winter-spring has ever been attempted, but data from various parts of the region have been collected. Figure 25 shows T-S plots of stations near St. Lawrence Island in April 1969 (*Staten Island*). Those taken east of St. Lawrence, closest to the Alaskan coast (41, 43), are the least saline, and station 50, near the central Gulf of Anadyr, has the highest salinity values. In addition to showing the existence in winter of the lateral salinity gradient, these data also illustrate the very well mixed water columns prevailing in winter,

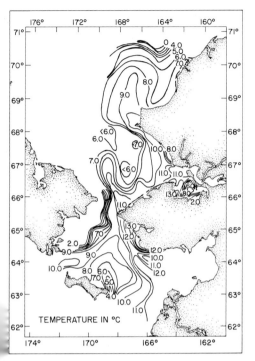 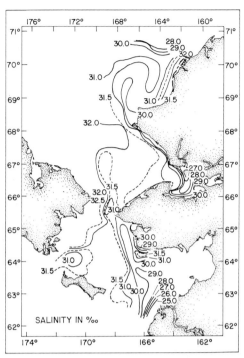

24. Distributions of temperature °C (left) and salinity ⁰/₀₀ (right) at 5 m observed by *Brown Bear*, 26 July–28 August 1960 (from Fleming and Heggarty 1966)

with temperatures close to the freezing point. The data strongly suggest that convection to the bottom is, or has been, taking place. The winter base condition for Alaskan Coastal Water is thus a fairly saline (31 to 32$°/_{00}$) bottom water with values increasing westward. The degree of dilution of this water depends directly on the quantities of fresh water available, and, therefore, on proximity to the coast.

In high latitudes, runoff varies enormously each season. Figure 26 shows the monthly mean discharge of the Yukon River measured at Ruby for selected years. Typically, discharge increases by an order of magnitude between April and June, and then decreases less rapidly toward the very low values of winter, so that there are inputs of fresh water exceeding 5×10^3 m^3 sec^{-1} through September.

The relatively abrupt addition of large amounts of fresh water into the region creates a sharp pycnocline, which is further reinforced by the rapid melting of the sea–ice cover from April to June. The pycnocline markedly stabilizes the water column, so that vertical convection ceases

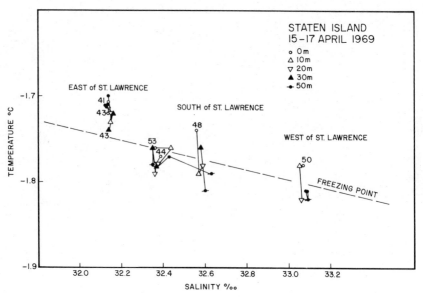

25. T-S diagram of stations east, south, and west of St. Lawrence Island, 15–17 April 1969 *Staten Island*

and the characteristic two-layer system of summer is created. The degree of stability developed is directly related to the quantities of fresh water introduced and proximity of sources. Figure 27 shows the distribution of maximum stability over a 10 m depth increment (calculated as $E = 10^{-3} (\Delta\sigma_t/\Delta z)$, Sverdrup et al. 1942, p. 417) for July 1968. The pattern is remarkably similar to that of upper layer salinity (Fig. 24b), with the highest stabilities associated with the lowest salinities and vice versa. The depth of the stability maximum was 10 to 15 m almost without exception, i.e., maximum $\Delta\sigma_t$ was located between 5 and 15m or 10 and 20 m.

A feature of the salinity regime of the region not elucidated earlier can now be seen. In addition to the very high stabilities associated with the Alaskan coastal runoff, apparent from the Yukon delta to Bering Strait, there is also a marked fresh water influence in and just north of the Strait of Anadyr. This must be attributed to sources of fresh water in the Gulf of Anadyr. Their effects are generally confined close to the coast and are not usually sampled in any of the U.S. oceanographic cruises. However, just north of the Strait of Anadyr the fresh water influence extends farther than 25 km eastward (cf. Figs. 24b, 27). This fresh water input from Siberia is much less than from Alaska. The mean annual discharge of the Anadyr River (1660 m^3 sec^{-1}) is only about one-fourth that of the Yukon (6220 m^3 sec^{-1}), and there are no other large Siberian rivers in the region

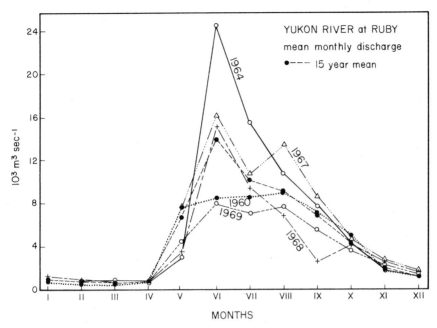

26. Monthly mean discharge of Yukon River measured at Ruby, Alaska for selected years and 15-year mean values

comparable to, for example, the Kuskokwim. Therefore, on the Siberian side the effects on the hydrography of the Bering Strait region are much less, nowhere near sufficient to give rise to a distinctly less saline water mass, and any effects are only observable locally.

The temperature distribution of the Alaskan Coastal Water in midsummer (Fig. 24a) also appears to be related to salinity. Fleming and Heggarty (1966) pointed out the general inverse relationship, in that the warmer surface waters are usually rather more dilute. They noted two exceptions: one was in the north, close to the ice, and the other was across the mouth of Norton Sound, where a tongue of relatively saline but warm water was defined offshore from Nome in 1960 (see Figs. 24a, b). The latter phenomenon was not defined in the 1968 *Staten Island* data, which covered the same area, and so does not appear to be general.

The inverse temperature-salinity correlation of the Alaskan Coastal Water surface layer suggests a relationship with stability. We have therefore plotted in Fig. 28 the upper layer temperature vs. maximum stability for the 1968 *Staten Island* data. The curve $10^8 E = 3.9 e^{0.42T}$ fits the data with a correlation coefficient $r = 0.93$, except for those stations

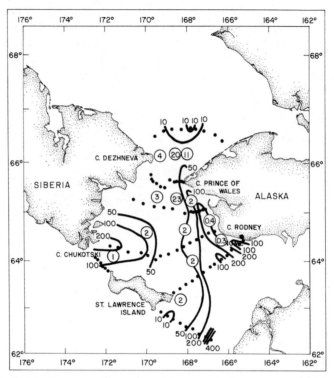

27. Distribution of maximum stability (10^8 E) in a 10-m depth increment, 1968 *Staten Island* (Circled numbers are vertical eddy coefficient K_V, cm^2 sec^{-1}, see Chap. 5.)

north of Bering Strait (see Fig. 3 section E). Comparison with the 1960 *Brown Bear* showed the following features:

(1) The exponential fit to *Brown Bear* data ($r = 0.88$) was similar in shape and, therefore, can be taken to be a usual feature of the midsummer thermal regime of the region;

(2) stabilities were generally higher in 1960, and there were no values of $10^8 E < 40$; and

(3) upper layer temperatures were ~2°C warmer throughout than those in 1968.

The differences noted are consistent with the differences in runoff affecting the region in the two years (see Fig. 26). Ruby, Alaska, where the Yukon River is gauged, is ~900 km from the delta. Water flowing at 50 cm sec^{-1} would require almost two months to reach the river mouth, while water entering downstream would require less time; some further time would be required for the water to spread at sea. Thus, in utilizing the

Yukon gauge data as an index of runoff influence, some time lag is required; we conservatively estimate one month. The 1968 *Staten Island* data were obtained mid-July and the 1960 *Brown Bear* data the end of July, so that the appropriate fresh water index is June river discharge. June 1960 discharge was about average, while June 1968 discharge was only one-half the average. Therefore, the seasonal and year-to-year changes in temperature, salinity and stability appear closely related to observed variations in runoff. (This is examined further in the section on *Year-to-Year Variations*.)

We shall now try to explain the observed thermal regime. The upper

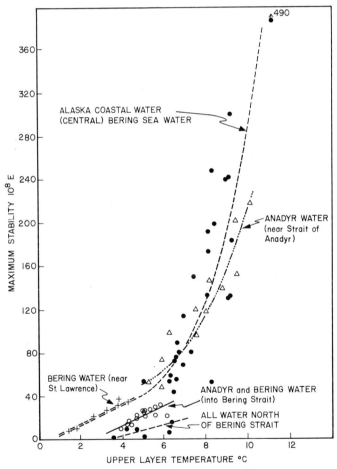

28. Plot of upper layer temperature °C versus maximum stability $10^8 E$, 1968 *Staten Island*

water layer in summer, the temperature of which varies by 10°C over the region, is subject to warming both through the addition of already warmed fresh water runoff and through radiant heating. The latter source provides approximately 3×10^2 ly day^{-1} during the period June to August in this latitude (Neumann and Pierson 1966, p. 241). If the warming were simply a matter of these processes acting over time on the surface layer, either no particular correlation between temperature and stability would be expected (since the runoff would vary widely in temperature), as would residence times of water with similar stabilities; or if there were a correlation, it would through mixing tend to be more nearly linear between cold and high salinity water and warm and low salinity water.

The highly correlated exponential relationship suggests a physically controlled thermal regime, as follows. The lower layer of the region is always colder (with local exceptions north of Bering Strait) and acts as a heat sink. The heat flux between layers is given by $\Delta Q/\Delta t = \rho C_P K_z (\Delta T/\Delta z)$ per unit area, where $\Delta Q/\Delta t$ is the heat flux, $\rho C_P \sim 1$ cal cm^{-3} °C^{-1}, K_z the eddy conductivity, and $\Delta T/\Delta z$ the temperature gradient. We note the important point that K_z is an inverse function of stability (see Proudman 1953). The sharp bend in the correlation curve (Fig. 28: in 1968 at $10^8 E \sim 40$) marks a separation of the regime into two parts. In the low stability region ($10^8 E < 40$), the temperature achieved by the upper layer is controlled by the rate of heat flux downward to the lower layer: thus, the higher the stability, the smaller both K_z and the heat flux, and the warmer the upper layer becomes. Temperature and stability values appear to be linearly related, with intercept values of temperature approximately those of the lower layer, near St. Lawrence Island $\sim$0°C and close to Bering Strait $\sim$2°C.

In the high stability region ($10^8 E > 40$), heat loss downward becomes so low due to suppression of K_z that it plays a relatively insignificant role in spite of a large $\Delta T/\Delta z$. The degree of warming achieved is governed largely by the temperature difference between the upper layer and the atmosphere; i.e., like a first-order rate reaction the upper layer gains heat at a rate proportional to the temperature difference, asymptotically approaching the upper limit imposed by the prevailing climatic conditions. There is a feedback in the process, because the warmer temperatures tend to increase the stability. We note that there is a similar relationship in the Anadyr Water near the Strait of Anadyr (cf. Fig. 28), where high stabilities are generated by the influence of Siberian runoff. These physical mechanisms governing summer temperature distributions in coastal waters would of course pertain to the development of water masses on continental shelves generally, wherever precipitation and runoff exceed evaporation.

To summarize, Alaskan Coastal Water is created from a relatively saline and cold water ubiquitous to the Bering Sea shelf, which water in turn represents a dilution by continental runoff of higher salinity water indigenous to the upper layers of the deep Bering Sea. The degree of dilution varies with proximity to the coast, so that the salinity increases away from the coast. Cold temperatures are generated in winter, when they become nearly uniform over most of the shelf. A strong pycnocline is created in late spring by an abrupt increase in runoff (greater than an order of magnitude in 1 to 2 months), and sufficient in volume to significantly reduce the salinities of the upper layer water over the whole eastern area of the region. This low-salinity water, which spreads over the surface in a relatively thin layer (order of 10 m), is advected downstream, and is identifiable far from the runoff source. Alaskan Coastal Water can be seen 100 km north of Bering Strait, even though the major fresh water sources are more than 200 km south of the strait. Only a very slow upward diffusion of salt is suggested.

With the advent in early summer of the less saline surface layer, temperatures increase. Where the salinity difference between upper and lower layers remains small and stabilities fairly low, bottom temperatures steadily increase through heat flux down from the upper layer, and in turn the upper layer temperatures remain relatively low. Where the salinity difference is great and likewise the stability, vertical heat (and probably also salt) flux is suppressed, heat can accumulate in the upper layer and its temperatures rise asymptotically toward the climatic limit. Thus, upper layer temperatures of this water mass by midsummer vary over a wide range ($\sim 10°C$).

The upper layer is destroyed during fall-winter by (1) the cessation of runoff, which by November is nearly down to midwinter values, thus removing the source of buoyancy to maintain the two-layer regime; and (2) cooling through loss of heat to the atmosphere, later followed by further extraction of fresh water through ice formation. Ice forms over the whole Bering Sea shelf in winter typically as far south as the Pribilof Islands (Lisitsyn 1966).

The stability-controlled thermal regime, demonstrated for the Alaskan Coastal Water but probably much more general in application, has important consequences for ice distribution. That is, development of the highly stable surface layer effectively allows concentration of the heat close to the surface, so that areas close to runoff sources become centers of ice disintegration in spring, and features of summer ice distributions over Arctic shelves are undoubtedly conditioned by the advection of such stratified coastal waters.

We comment on Norton Sound separately. This embayment of the system has never been surveyed; only a few widely spaced stations were taken from the *Northland* (Goodman et al. 1942). Fleming and Heggarty (1966) reported that fresh water, primarily from the Yukon River, tends to accumulate here in summer, and they deduced from drift bottle studies that circulation is cyclonic. They also remarked on relatively warm and saline water close to shore off Nome which they felt was issuing from Norton Sound (Fig. 24). Current measurements at 5 m and 20 m in this water showed westerly flow.

We find no evidence for warm and more saline water off Nome in the 1968 *Staten Island* section. At that time, all water in the northeastern part of the section (section A', Fig. 3), some of which was undoubtedly coming from Norton Sound, was characteristic of Alaskan Coastal Water, and was indistinguishable from coastal water south of the Yukon delta (section A, Fig. 3). In the June 1969 *Northwind* section (Husby and Hufford 1971), the 31°/$_{oo}$ isohaline lay 5 m shallower over the channel, which extends eastward in Norton Sound, than it did farther west, but in other respects the water properties and gradients were typical of Alaskan Coastal Water.

The occasional presence of relatively dense water, over a submarine canyon at shallow depths, where the upper layer flow is down-canyon, may be similar to the regime over Svataya Anna, the large submarine canyon which indents the northern Kara Sea (Coachman and Barnes 1962). There the canyon appears to play a dynamical role not unlike a sunken estuary, in which surface water flows outward and the more saline deeper water flows inward and is raised to shallower depths. A similar phenomenon may occur, at least occasionally, over the channel off Nome, and thereby account for the 1960 *Brown Bear* observations. Data are insufficient to be more conclusive.

We believe that the primary importance of Norton Sound is to provide an extensive region for the formation of Alaskan Coastal Water. It may however have a secondary role in providing particular conditions for the generation of local anomalies in water mass properties, but this awaits confirmation.

LONG-TERM TEMPORAL VARIATIONS

We have mentioned that there are more or less regular variations in water mass properties with season, and the annual cycles of temperature and salinity in Alaskan coastal water have been discussed in some detail. There are also detectable year-to-year variations.

Seasonal Changes

Major seasonal variations in water mass properties are related to (1) the large quantities of fresh water abruptly introduced from May to June, which in somewhat lesser amounts continue to enter the sea into September before tapering off to the extremely low values of winter (Fig. 26), and (2) a marked variation in the net surface heat exchange available for heating (or cooling) surface layers. In the Bering Strait region, the radiation balance is that of a net heat loss from approximately mid-September to mid-March. From about the first of April to September, the balance is such as to provide heating, with a peak in July (Marshunova 1961; Fletcher 1965).

To examine the winter condition, when the water columns throughout the region are apparently nearly isothermal and isohaline, numerous high-quality observations (67 stations) are available from the 1969 *Staten Island* cruise. There were two phases to this cruise, the first, from 14 February to 3 March, occupied 20 stations east and northeast of St. Lawrence, and the second phase, from 5 to 25 April, occupied 47 stations east, south, and west of St. Lawrence, with a traverse northward through the eastern channel of Bering Strait.

A frequency distribution of the deviation of the temperature from the freezing point at 1 atmosphere pressure for the observed salinity for all observations, is presented in Fig. 29. The previous conclusion about the water at all depths being maintained close to freezing in winter by the physics of the freezing process seems conclusively verified.

The data displayed in Fig. 29, however, give further insight into the processes obtaining. The data from the two cruise phases fall into separate populations; the midwinter phase has a mean deviation from the freezing point of $-0.02°C$ and the late-winter phase $+0.005°C$, which are significantly different at the $>99.9\%$ level. Furthermore, the variances of the two populations differ, that of midwinter being less than that of late winter, significant at the $>80\%$ level. The instuments and scientific personnel (R. B. Tripp, cruise leader) were the same for both phases. Temperatures were measured by carefully calibrated reversing thermometers and the accuracy is considered to be better than $\pm 0.015°C$. The salinities were run on a Bissett-Berman salinometer calibrated against Copenhagen water and are considered accurate within less than $\pm 0.01°/_{00}$. The relationship used for calculating the freezing point depression at 1 atmosphere was that of Thompson (Sverdrup et al. 1942, p. 66). Thus, the water column on the average in midwinter is slightly supercooled relative to surface pressure, while in late winter it is on the average at the freezing point. Data from 13 stations taken from *Northwind* 15 March to 4 April 1955 east and northeast of St. Lawrence show the same distribution about freezing as the late-winter *Staten Island* data.

We interpret these findings as follows. During midwinter, when there is a large heat flux to the atmosphere, heat is continually being subtracted from the water, abetted by a vigorous vertical convection, and the

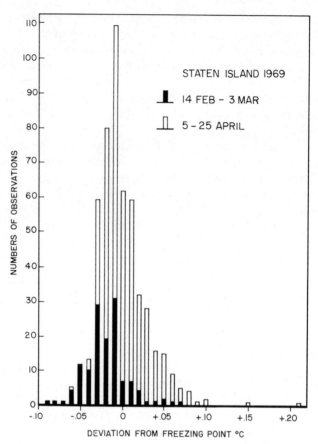

29. Deviations from freezing point during midwinter (14 February–3 March) and late winter (5–25 April) observed by *Staten Island* in 1969

temperatures pulled below freezing by the high cooling rate. The water at depth may not be actually supercooled *in situ* because of the higher pressure, but occasional supercooled values are encountered near-surface, and as supercooled water is metastable in the presence of ice (Lewis and Lake 1971), freezing must be actively occurring.

In late winter the conditions are: (a) reduced heat flux to the atmosphere (in fact slight warming may have begun in certain instances); (b) an insignificant amount of freezing; and finally (c) ice and water in general coexisting at the equilibrium temperature (freezing point). The lesser variance of the midwinter temperatures suggests that when freezing is active the range of temperatures is more severely restricted by the physics of freezing, while in late winter when freezing is not active, warmer water (i.e., slightly above freezing) can persist in the column if it is away from the ice (presumably there is less vigorous convection to bring it into contact with the ice).

To elucidate the changes in temperature and salinity during the seasonal cycle, average cross-sections of Bering Strait were constructed for early summer, 5 July to 7 August, and late summer, 20 August to 5 October (see Table 1 for data sources). The averaging procedure is necessary to suppress year-to-year and short-period variations. Few data are available from other times of the year. There is one detailed cross-section from June 1969 (Husby and Hufford 1971) and the few stations from the eastern channel of Bering Strait taken from the *Staten Island* in April 1969. Also, a section was constructed from the *Northwind* data of 26 October 1962. Though the stations were not well placed, cross-strait variations were small and the values are probably representative of fall. Two of the April 1969 stations are presented in Fig. 30, the average cross-sections in Figs. 31a, b, and the fall section in Fig. 32.

The vertical distribution of salinity in Bering Strait in late April 1969 showed the beginning of a two-layered structure, with a sharp halocline (therefore pycnocline) at depths of 37 m and 27 m at, respectively, 11 and 15 km west from Cape Prince of Wales. The layering may have two dif-

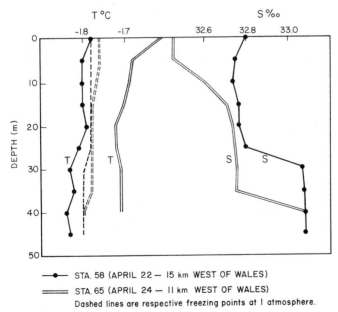

30. Temperature °C and salinity ‰ for two stations from Bering Strait, April 1969 *Staten Island*

48 • BERING STRAIT: The Regional Physical Oceanography

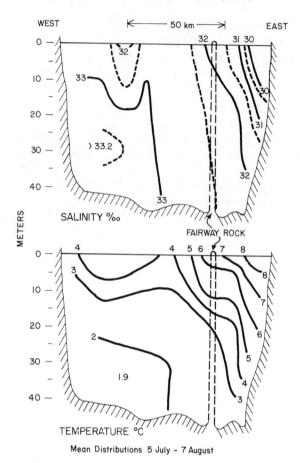

31a. Average early summer (5 July–7 August) salinity and temperature cross-sections of Bering Strait

ferent origins. First, though layering in the water columns is not observed upstream (south) of the strait in winter, there is a general gradient of salinity increasing westward across the region, and adjacent water columns of differing salinities, on being advected into the strait, might tend to layer, so that even in winter isohalines within Bering Strait would slope upward to the west. If the situation had not changed materially in the two-day interval separating these stations, the attitude of the 33‰ April isohaline would be the same as that of the 31.6‰ early summer isohaline (Fig. 31a).

Second, as of late April 1969 fresh water runoff from the Alaskan shore had begun in quantities sufficient to be reflected in the regional salinity

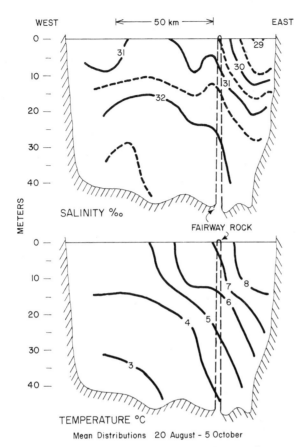

31b. Average late summer (20 August–5 October) salinity and temperature cross-sections of Bering Strait

distributions. The Yukon River gauge data showed that the marked spring increase in runoff began during April 1969. Furthermore, there was a secondary shallow halocline at the station closest to Alaska (65) suggestive of a fresh admixture, and the water had been warmed to ~0.1°C above freezing.

We cannot definitely choose between these alternatives, but feel that the shallow halocline probably represents the onset of water mass property variations in the spring, while the deep halocline is a permanent feature resulting from transformation of the lateral salinity gradient south of the strait into a vertical layering. Then, as summer progresses, more and more fresh water accumulates in the system. This is clearly seen by examining the average salinity sections (Figs. 31a, b) in which the

50 • BERING STRAIT: The Regional Physical Oceanography

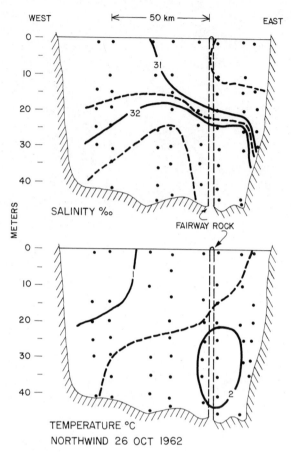

32. Salinity and temperature cross-section from Bering Strait, 26 October 1962 *Northwind*

isohalines are displaced progressively westward across the strait with time.

The seasonal change in fresh water content is further illustrated in Fig. 33, which plots the approximate positions of $32°/_{00}$ and $33°/_{00}$ isohalines in the lower water layer at various times in the seasonal cycle. All cruises which covered a substantial portion of the region were utilized and the data smoothed to suppress year-to-year and short-period variations. The regular progression of the isohalines westward from summer into fall and their return eastward by late winter is evident.

Thus, a major seasonal variation involves the accumulation of fresh water and consequent reduction of salinities over the summer, when runoff is high, followed during fall/winter by the fresh water being

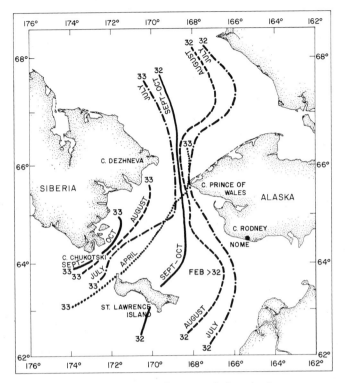

33. Approximate locations of 32°/oo and 33°/oo isohalines in deep water at various times during seasonal cycle, smoothed, from all available data

flushed out of the region. Order of magnitude estimates were made of these changes as follows. For the Bering Strait cross-section, an equivalent area of fresh water was determined by

$$\text{F.W. Area} = \frac{S_o A^* - \int_0^{A*} S dA}{S_o}$$

where S = salinity, A = area, S_o = base salinity (approximately 33°/oo, cf. Fig. 13b), and A^* = area of $S < S_o$. This is equivalent to the fresh water-fraction calculations of Ketchum and Keen (1955) and Tully and Barber (1960). As the mean flow throughout the region is north (see Chap. 3, Currents), fresh water introduced south of the strait is removed from the system as part of the Bering Strait transport. We now assume the equivalent area of fresh water determined in Bering Strait to represent the fresh water fraction over a region subtending ~4° of latitude = 450 km. The balance of fresh water in the system, neglecting precipitation, is given by

Accumulation Rate = Runoff − Transport.

Calculations were made for the mean salinity sections of early and late summer (Figs. 31a, b) from the sections available from June (*Northwind* 1969) and the end of October (*Northwind* 1962, Fig. 32) and the late April 1969 *Staten Island* data.

The results are presented in Table 3. Although the values should not be regarded as precise, the order of magnitude of the transport and runoff appears reasonable and provides the following insights:

(1) The fresh water accumulation rate increases in early summer along with the runoff and melting of sea ice, and it peaks in June/July. The rate becomes small but is still positive through late summer into fall, when it becomes negative.

(2) The quantity of fresh water in the system is phased later, the peak in fresh water accumulation occurring in late summer. At the end of October there is still a great deal of fresh water in the system. It appears to take about two months after the runoff has dropped to low levels to flush the accumulated fresh water from the system (i.e., late December to early January).

(3) Yukon River flow values at Ruby would seem to represent 10 to 15% of the total fresh water influx. [Note that Ruby is ~900 km from the mouth of the Yukon, where the flow is over 3 times greater than at Ruby].

There is a seasonal cycle in water temperature which goes hand in hand with the fresh water cycle. Examination of the average sections (Figs. 31a, b) shows the maximum temperature (>8°C in these sections) to be achieved by late July, shortly after the fresh water accumulation has achieved its largest increases. The maximum temperature remains nearly the same through late summer. The cycle is shown graphically in the recorded temperature data reported by Bloom (1964, Fig. 34). The temperatures of Alaskan Coastal water are seen to have risen steadily and peaked by late July and then remained high until the latter part of September. The cause for the temperature plateau of summer lies in the physics of the warming process, which sets a limit on maximum temperature as discussed earlier.

Even though maximum temperatures are achieved by July, the total heat content continues to increase in the system through late summer, as the higher temperatures are observed farther west across the strait in the late-summer section (Fig. 31b) than in early summer (Fig. 31a).

Fedorova and Yankina (1964) published values of monthly mean temperature of the Bering Strait cross-section, reproduced in Table 4. Also included are our values, from the sections cited above.

TABLE 3
Components of Seasonal Freshwater Cycle

	Freshwater Area 10^3 m^2	Accumulated Volume (L=450 km) 10^9 m^3	Days	Accumulation Rate 10^3 m^3 sec^{-1}	Transport[a] 10^3 m^3 sec^{-1}
April	6	2.7			
			55	4.2	14
June	50	22.5			
			30	16.7	49
July	146	65.7			
			50	1.5	76
September	160	72			
			45	−1.0	78
October	153	69			
			$T = 55$ (calculated from freshwater balance, see below)	$\left[-\frac{69}{T}\right]$	38
Winter	0	0			

		————10^3 m^3 sec^{-1}————			
Period	Accum. Rate	Transport	Runoff	Yukon R.[b] at Ruby	Factor[c]
Apr–Jun	4	14	18	2.5	7.2
Jun–Jul	17	49	66	10	6.6
Jul–Sep	2	76	78	10	7.8
Sep–Oct	−1	78	77	8	9.6
Oct–Winter	$\left[-\frac{69}{T}\right]$	38	(3×7.8)	3	mean = 7.8

[a] Freshwater Area × mean speed (=50 cm sec^{-1}).
[b] Average of prior 2 months (Fig. 26).
[c] To multiply Yukon River (Ruby) to represent runoff.

Values from the present study are in close agreement with the Soviet data in fall and winter. In the summer, however, they are about 1°C greater. A possible explanation is that, as the warmest water always lies along the Alaskan coast, the Soviet sampling may have been too sparse in the eastern part of the section (Fedorova and Yankina did not report data sources), and thus their value could be biased toward the lower temperatures of the western side. Alternatively, it could be that as the Soviet data were acquired before 1960 while the bulk of our data after 1960, the marked warming observed in the 1960s (see below) is not reflected in the Soviet results.

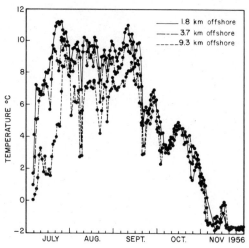

34. Temperatures near-bottom recorded at various distances west of Wales, Alaska, July–November 1956 (from Bloom 1964)

TABLE 4
Monthly Mean Temperature (°C) of Bering Strait Section

Month	I	II	III	IV	V
Fedorova and Yankina	−1.8	−1.8	−1.8	−1.8	−1.5
Present Study				−1.8	

VI	VII	VIII	IX	X	XI	XII
0.2	2.1	3.4	4.0	1.8	−0.05	−1.8
0.5	3.3		4.8	1.5		

In summary, the seasonal cycle of heat in the system closely parallels that of fresh water accumulation (cf. Table 3), with a buildup to high values by July, and then a less rapid but continuing increase to peak values in September. Fall cooling through the sea surface results in the warmer water being found at depth during that season (Fig. 32). Small amounts of ice formation may take place locally as early as October, but significant ice formation would not begin until the accumulated heat is flushed from the system, which is probably during November or December.

Year-to-Year and Long-Term Variations

We have noted that the water masses in the Bering Strait region are defined by salinity bands, but that the median salinity of the bands differs from year to year (Table 2). We hypothesize that as runoff and quantities of fresh water input can differ markedly year to year (Fig. 26), the degree of dilution and hence salinity range of the water masses varies commensurately.

To investigate this, Fig. 35 shows gauge data of the Yukon River at Ruby and estimates of median values and ranges of salinities of the water masses as functions of time. While the gauge data are insufficient to allow precise estimates of total runoff into the system, they at least give a reasonable index of whether runoff is above or below average. Hence, the river flow is plotted as deviations from the appropriate average mean monthly flow, and deviations for the 6 months of high discharge (May to October) are also shown.

The salinity values for the water masses were taken from T-S diagrams of all stations in, or close to, Bering Strait. There is subjective judgement in this procedure. The coverage of the strait provided by some cruises, particularly the earlier ones, was poor; and in some instances the selection of boundaries separating water masses is somewhat arbitrary (cf.

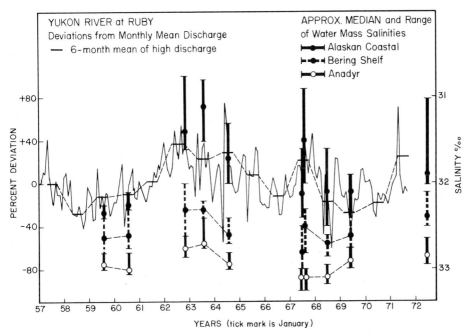

35. Deviations (%) of monthly mean discharge of Yukon River at Ruby from 15-year monthly means, 1957–1971, and approximate median and range of water mass salinities from various cruises 1959–1972. (Bars are deviations from means of six highest discharge months, May through October.)

Figs. 4 to 7). Frequently, observations near the surface (<10 m) have been ignored.

There are definite variations in runoff from Alaska, with anywhere from 1 to 4 years separating times of low and high discharge. A comparison of observed extremes (Table 5) shows that the quantity of fresh water introduced annually can vary by a factor of 2.

TABLE 5
Extremes of Alaska Coast Runoff (1957–1971)

	Mean Annual Flow Rate (Yukon River at Ruby) 10^3 m^3 sec^{-1}	Total Annual Volume (Yukon × 8) km^3 yr^{-1}
High	6.5 (1962)	1.6
Low	3.5 (1958)	0.9

It appears (Fig. 35) that the salinities of the water masses on the whole reflect the variations in runoff: in the years of above-average runoff salinities were low (1962 to 1964) and vice versa (1959 to 1960). The Alaskan Coastal Water shows the greatest variations (up to about 1°/$_{00}$) as well as the widest range of values, because it is the water mass most directly and rapidly affected by the coastal discharge. The Bering Shelf and Anadyr waters show lesser variations (0.3 to 0.4°/$_{00}$ between extremes) because the effect of the discharge is rather indirect: the runoff must first spread over large regions of the shelf and mix with the deeper waters.

We can check whether the difference in volume of fresh water introduced would be sufficient to cause the observed change in salinity, as follows. The shelf area influenced by discharge is assumed to be about 4° latitude × 7° longitude, the mean depth 50 m, and the water in this volume of salinity 32.6°/$_{00}$. The discharge rate required to reduce the salinity of this volume 0.5°/$_{00}$ during the four months June to September would be 15 × 10^3 m^3 sec^{-1}, or approximately 1.5 times the mean flow rate of the Yukon River at Ruby during the corresponding months. Since the discharge measured at Ruby represents only 10 to 15% of the total runoff (Table 3), and the summer discharge rate can vary by a factor of 2 (cf. Fig. 26), there is adequate variation in the fresh water supply to alter the salinities as observed, subject to the residence time of the water in the specified shelf area being in the neighborhood of 30 to 45 days. This would in turn require a minimum average flow of 11 to 17 cm sec^{-1}, which is probably reasonable (cf. Figs. 46, 47).

The highest salinities ever reported for Bering Strait were in 1932 and 1933 (Ratmonov 1937a, b), when the ranges for Anadyr Water were about 33 to 33.25°/$_{00}$ and 33.2 to 33.45°/$_{00}$, respectively. In 1934 the water mass was much less saline (32.6 to 32.95°/₀₀). In 1937 the salinities ranged between 33.15 and 33.45°/₀₀, followed in 1938 by 32.75 to 33°/₀₀ (*Northland*, Goodman et al. 1942). No discharge data are available from that time, precluding a comparison with available fresh water supply.

During only two years in the recent decade have salinity ranges comparable to those in the 1930s been observed for the Anadyr Water (1967: 33 to 33.25°/₀₀; 1968: 32.95 to 33.15°/₀₀, Fig. 35). It is possible that the Soviet salinity determinations of 1932 and 1933 are not comparable with the U.S. observations; however, the 1937 *Northland* results were from the University of Washington Oceanographic Laboratories directed by T. G. Thompson.

These relationships are depicted in Fig. 36, showing year by year the estimated salinity range of Anadyr Water. The 0.2°/₀₀ difference in mean salinity between the two year groupings is significant at the 5% level, using the student's t-distribution. We tentatively conclude that there has been a long-term decrease of about 0.2°/₀₀ in 30 years of the general salinity of the Anadyr Water, on which trend the shorter-term variations are superposed.

Trends in the temperatures of water masses transiting Bering Strait were investigated by planimetering detailed temperature cross-sections to obtain mean values. The mean temperatures together with maximum

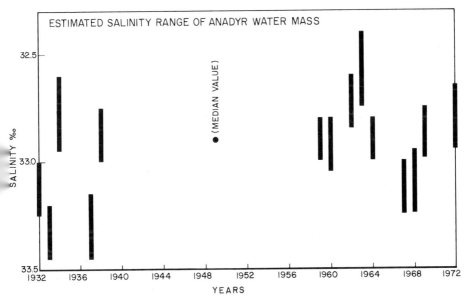

36. Estimated salinity range of Anadyr water mass in various years 1932 through 1972

and minimum observed values and northward transport are presented in Table 6.

One major trend can be detected, viz., the water transiting Bering Strait in midsummer in recent years has been about 1°C warmer than in the 1930s. Figure 37 presents the mean temperatures together with the seasonal cycle (Table 4), and with seasonal adjustment the cold temperatures observed in the period 1932 to 1934 stand out in particularly sharp contrast to the warmer temperatures observed since 1960.

It is possible that there is another long-term variation affecting temperature values in the Bering Strait region, though data are so sparse and short-term variations so large (cf. 1967 data in Table 6) it cannot be convincingly demonstrated. Nevertheless, the tendency we see is that with greater quantities of fresh water input the mean temperatures tend to be lower. Thus, 1964 and 1972 were years of above-average discharge and

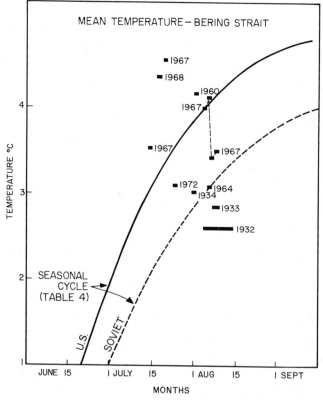

37. Mean temperature in Bering Strait during July and August of various years, 1932–72: curves of seasonal temperature trend from data of Table 4

TABLE 6
Midsummer Temperatures and North Transport, Bering Strait Section

Date		Observed Temps. (°C)		Mean Temp. °C	North Transport Sv.
		Maximum	Minimum		
1932	3–14 Aug	8.8	−0.3	2.59	
1933	6–9 Aug	11.6	0	2.82	
1934	31 Jul–1 Aug	7.5	1.0	3.00	
1937	5–6 Jul	*	−0.2		
1938	22–23 Aug	*	1.4		
1959	2 Aug	*	2.4		
1960	1–2 Aug	10.1	1.3	4.15	
1963	7–8 Aug	*	2.6		
1964	5–6 Aug	10.6	1.0	3.06	1.4
1967	13–14 Jul	7.6	2.0	3.51	2.0**
	18–19 Jul	9.3	1.7	4.54	−0.2
	3–4 Aug	9.5	1.6	3.99	2.2
	4–5 Aug	10.3	1.5	4.10	1.6
	5–6 Aug	13.9	1.5	3.40	2.0
	6–7 Aug	10.9	1.1	3.48	1.4
1968	17–18 Jul	8.8	2.2	4.35	1.5
1972	24–25 Jul	13.7	1.1	3.08	1.7

*Undoubtedly the warmest water in the Strait was not sampled.
**Extrapolated (see Chap. 3, Currents).

the salinity ranges of the Alaskan Coastal Water were low (cf. Fig. 6) while the mean strait temperatures were the lowest observed since 1960. Conversely, 1960 and 1968 had below-normal fresh water supply and the mean cross-section temperatures were among the highest observed. We point out that it is not the temperature values per se which seem to be correlated in this way, but rather the integrated values (i.e., the heat content).

A possible explanation of such a correlation is the insulating effect of the fresher surface layer. In years of above-normal discharge the stability is greater, which markedly reduces the flux of heat downward to the lower layers. Then, as the warmth of the upper layer is limited by the climatic air temperature, the net effect is that less total heat can be added to the system and the mean temperature is lower.

SHORT-TERM VARIATIONS

Significant local change in water properties can occur on time scales from less than one day to 1 to 2 weeks, and in this section we investigate such variations.

Time Scales of Less Than One Day

Data for assessing the most rapid changes are very scarce, consisting of four anchor stations of one-day duration with observations every 2 to 5 hours (*Chelan* 1934 station 45, located just south of the western channel of Bering Strait; and three stations from *Staten Island* 1968, Fig. 3: stations 1XX, 2XX, 3XX).

Temperatures and salinities observed at 0 m and 25 m at the four stations are plotted in Fig. 38a, b. These depths are taken as representative of variability in the upper and lower layers, respectively. Variability in the lower layer is much less than in the upper. The changes in the lower layer over one day are in general <0.3°C and <0.05°/$_{00}$, and appear to be slow drifts of the values either up or down. Station 2XX showed a definite drift down of temperature of 0.5°C but no salinity change. Station 1XX showed the only rapid marked change, an increase in temperature of 0.5°C between hours 8 and 10, while the corresponding salinity change was slight. At all these stations current speeds were substantial, averaging between 20 and 60 cm sec^{-1}.

Our interpretation is that either the deeper layer is exceedingly homogeneous within each water mass [the current speeds indicate that many kilometers of water passed each station each day (24 hrs × 20 cm sec^{-1} = 17 km)], so that the small drifts in values are due to the advection of water with low property gradients; and/or the flow is primarily along isopleths. Gradients across the boundaries between water masses are quite sharp [cf. T-S profiles from adjacent stations in Figs. 4 through 7; e.g., stations 57 and 58 (Fig. 6) were 5.5 km apart and showed 1°C different temperatures in the deep water]. As the boundaries must shift position somewhat with time, an occasional steep gradient can be advected past any particular location, causing a change such as observed at station 1XX.

In the upper layer, local changes in both temperature and salinity are large (in the present records as great as 3°C and 0.6°/$_{00}$). The variations are not systematic in any discernible way — the changes are clearly not periodic on time scales <1 day, nor are the T-S correlations constant. Sometimes salinity and temperature vary in the same sense and sometimes oppositely. We envision the upper layer to have embedded numerous puddles or lenses of water with characteristics different from ambient, the puddles being advected through the system. These puddles must be numerous and thus a key feature of the upper layer, for they are evidenced at all four stations. The measured current directions associated with a property change at each station are also plotted in Figs. 38a, b. Frequently the flow altered direction between the times of maximum and minimum T-S values (for stations *Chelan* 45 and 2XX it was

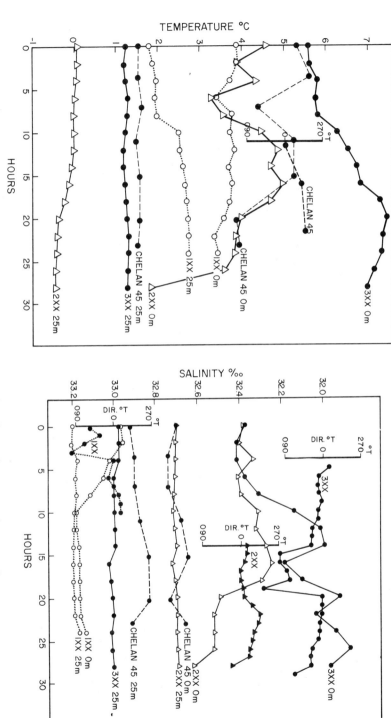

38. Temperatures (left) and salinities (right) observed at 0 m and 25 m at four one-day anchor stations, and current directions measured concurrently at three of the stations

about 45°, while for stations 1XX and 3XX about 90°), but no simple interpretation of these events as more or less regular or symmetric eddies appears possible from the present data. However, it seems clear to us that interpretation of the variations due to puddles (rather than frontal movement between relatively homogeneous water masses) past the meters is preferable. For example, under the latter interpretation the changes in temperature and salinity would be systematic according to the T-S correlations.

The size of the puddles or lenses can be estimated. For example: station 2XX showed a change from high salinity and low temperature water at about hour 6; to high temperature and low salinity water, with the extreme at hour 16; the change then reversed toward colder, more saline water. The mean speed during the time was 40 cm sec^{-1}, and the minimum lens half-length in the direction of flow was about 14 km. The estimates are summarized in Table 7.

TABLE 7
Estimates of Lens Size in Direction of Flow

Station	Chelan 45	1XX	2XX	3XX*
Avg. time of max. change, hrs	6.5	8	10	4.5
Mean speed cm sec^{-1}	22	9	40	47
Minimum half-length, km	5	3	14	8

*Based on salinities only.

It appears that the characteristic size of the lenses in the upper layer is perhaps 10 to 30 km across.

We conclude that a common feature of the upper layer is pools of water with different values of temperature and/or salinity, and sized a few 10s of kms across, being advected through the system. Thus, marked local changes of near-surface values, with a time scale of one-half day or less, are anticipated during midsummer when wide ranges of temperature and salinity obtain.

We recognize also that T-S observations from only one or two individual stations could give misleading results when used to interpret average conditions. When numerous stations in a section or grid much larger overall than the lens size are used, the variations would tend to cancel. Therefore, while caution in interpreting data from individual stations is necessary, the results reported here should not be seriously aliased, because we have mainly used sections with numerous closely spaced stations. Certainly, a truly synoptic picture of the upper layer temperature distribution cannot be presented. For example, the distributions of Fig. 24a, b (5 m T and S) are smoothed results.

Time Scales Longer Than One Day

Changes in water properties occurring over a few days can be evaluated from the 1967 *Thompson* data, when four detailed Bering Strait sections were made in one week. Pertinent parameters along the four sections, as well as along the two sections made about two and three weeks earlier (*Northwind*, 13 to 14 July; 18 to 19 July), are gathered in Table 8: the mean values of temperature, salinity, and transport; the percent of cross-sectional area occupied by three ranges of salinity; and maximum and minimum observed values of salinity with their corresponding temperatures. The salinity bands correspond approximately to the salinity ranges of the three major water masses (cf. Fig. 5).

While there generally are quite large variations on short time scales, by far the largest is the one on 18 to 19 July when the net transport had completely reversed from the normal northerly flow to southerly. Because southward transport in the system has so rarely been documented, we consider it an anomalous event and the accompanying T-S changes are discussed in the next section.

When the water in the system is flowing northward in the normal manner (transport $> +1$ Sv) the T-S variations day to day and week to week are less, and in large part seem to be systematic. A major variation is due to the seasonal trend. There is, in salinity, practically no change in the higher values, but a definite decrease in the lowest observed values (min. down from $30.9°/_{00}$ on 14 July to $26.7°/_{00}$ on 7 August). This reflects the increase in fresh water in the system over the summer, mainly in the Alaskan Coastal Water. Because the fraction of the strait section occupied by Alaskan Coastal Water is, during normal conditions, so small ($<12\%$ in these data), there is only very little effect on the average salinity. Also, there is practically no change in maximum salinity. The latter is due to the homogeneity of the Anadyr Water which experiences very little salinity alteration during its northward passage (cf. Fig. 16).

The temperatures also reflect seasonal change. The temperature of the minimum salinity (Alaskan Coastal Water) rises from $7°C$ to $10.9°C$, while that of the maximum salinity (Anadyr Water) showed a small but steady decrease of about $0.8°C$. The warming of the Alaskan Coastal Water until the midsummer thermal plateau has already been discussed. The cooling of the Anadyr Water is somewhat less obvious in origin, but may reflect a more effective mixing with water from the cold central Gulf of Anadyr as summer progresses. The decrease is regular enough to appear to be more than coincidental.

Certain variations also appear to be associated with variations in transport. For example, Fig. 39a suggests that the area of the section occupied by water with $S <32°/_{00}$ changes systematically with transport, the Alaskan

Coastal Water portion being of increasing relative importance with decreasing northerly transport. Thus, the water mass boundaries seem to shift westward across the strait during reduced north flow. In the one case of south flow, it is possible that water of characteristics similar to Alaskan Coastal Water, which comes from Kotzebue Sound and moves southwest along the coast toward Shishmaref (see Chap. 4, *Chukchi Sea*), has also moved into the strait.

During the first week of August there was also a change in the character of one or more of the water masses, as suggested in Fig. 39b. If the

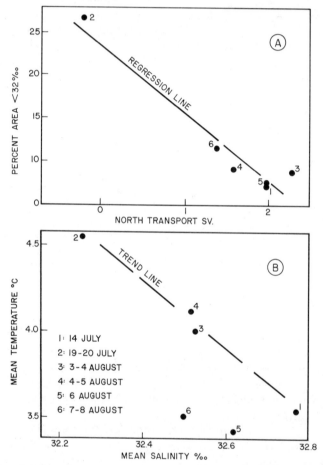

39. (A) correlation of percent area of Bering Strait occupied by water with $S < 32^0/_{00}$ with north transport (Sv = Sverdrup); (B) mean temperature-salinity characteristics for Bering Strait for the six detailed sections occupied between 14 July and 8 August, 1967

TABLE 8
Bering Strait Section—Summer 1967

	Mean Values			Area %			Observed Values			
	T °C	S ‰	Tr., Sv	>32.8	32–32.8	<32	Max. S ‰	Associated T °C	Min. S ‰	Associated T °C
13–14 July	3.51	32.77	2.0	65.5	27.5	7.0	33.23	2.46	30.92	7.07
18–19 July	4.54	32.26	−0.2	42.3	31.2	26.5	33.18	2.33	29.36	9.31
3–4 Aug.	3.99	32.53	2.3	60.8	30.4	8.8	33.25	1.98	27.66	9.52
4–5 Aug.	4.10	32.52	1.6	60.6	30.4	9.0	33.23	1.77	27.52	10.30
5–6 Aug.	3.40	32.62	2.0	62.0	30.5	7.5	33.13	1.71	27.34	10.29
6–7 Aug.	3.48	32.50	1.4	62.9	25.8	11.3	33.22	1.65	26.71	10.93

major water masses remained relatively constant in temperature and salinity, then the mean T-S point should vary nearly along a line connecting the mean T-S point of each water mass. However, the observations of 5 to 7 August show mean temperatures anomalously low by about 0.5°C. We assume that the mean salinities are not also anomalous, but can offer no proof, other than that there was little change in the quantities of each water mass present, as determined from the salinity ranges (cf. Table 8). As the Alaskan Coastal Water increased in temperature during this interval (cf. Table 8), the mean decrease must be due to decreases in the other two water masses. Some decrease in the Anadyr Water temperatures is indicated by Table 8, but not sufficiently large to account for an overall drop of 0.5°C. Therefore the Bering Shelf Water must have dropped in temperature on the order of 1 to 2°C. We cannot meaningfully pursue the matter further on the basis of available data, but rather cite this instance as an illustration of the considerable short-term variations that can occur.

ANOMALOUS EVENTS

19 to 20 July 1967

Major deviations of the flow through Bering Strait from its generally northward course with transports in excess of one Sv are rare. Coachman and Aagaard (1966), in an analysis of observations to 1964, concluded that southerly currents were occasionally observed locally in the surface layers and in the deeper water in the western channel, but that net southerly transport had never been documented. We now have one case in which we know there was net south transport, viz., 18 to 19 July 1967. One week prior (14 July) and two weeks following (3 to 4 August) the net transport was north at the normal rate. These major changes in flow are discussed more fully in Chap. 3.

The longest series of recorded currents in Bering Strait were reported by Bloom (1964). Uncertainties in the measurements (electric cables and acoustic meters) and calibrations preclude utilization of the values to estimate transport in the strait (cf. Coachman and Aagaard 1966). However, the relative changes in the data may suggest periods when the flow might have reversed, i.e., they indicate the times when south transport might conceivably have occurred. Figure 40 shows the incidence and duration of such periods in the interval September 1956 to August 1957, from Bloom's record. Thirteen incidences occurred during the year, concentrated in the fall and lasting between 1 and 9 days (but most commonly 2 to 5 days).

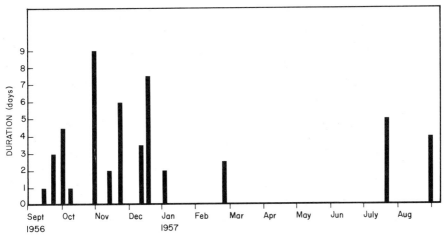

40. Incidence and duration of possible south flow in Bering Strait from geopotential measurements between September 1956 and August 1957 (data of Bloom 1964)

Figures 41 and 42 show T-S diagrams of the sections before and during the south transport of July 1967. On 14 July (Fig. 41) the T-S distribution was normal, with the three major water masses clearly distinguishable [this was also the case for the section of 3 to 4 August (Fig. 5), following the southerly flow event]. Table 8 data show the ratio of Anadyr: Bering Sea: Alaskan Coastal Water to have been the normal 6:3:1. During the event (Fig. 42), the water mass distributions were markedly different. Anadyr water appeared to be unchanged in character but was represented at fewer stations. The boundary between Bering Shelf and Alaskan Coastal Water changed character, so that instead of the normal more or less vertical water mass separation, the intermediate station (61) showed a layering of a relatively cold low-salinity component from the Alaskan Coastal Water overriding the Bering Shelf Water. Furthermore, Alaskan Coastal Water was represented at more stations than usual.

This latter change is also apparent in Table 8, showing the water mass ratio to have changed to 4:3:3. (The distinction between Alaskan Coastal Water and Bering Shelf Water was somewhat blurred by mixing.) In Fig. 43, the fresh water area (see *Seasonal Change*) is plotted for the six 1967 sections, and the trend of increase in fresh water (Table 3) over this season is noted by the dashed line. On 18 to 19 July, however, the fresh water area was doubled, representing a change one-half order of magnitude more than that which occurs during the normal short-period variations (cf. data for 3 to 7 August).

The maximum and minimum values of temperature and salinity were little altered during the event. The mean values in the section changed in

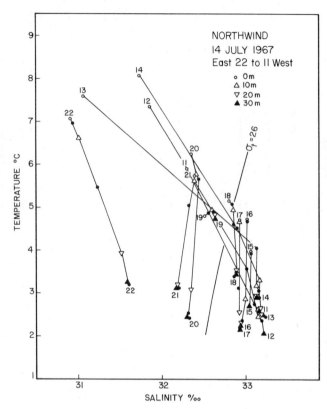

41. T-S diagram of Bering Strait section, *Northwind*, 14 July 1967: station numbers in order from east to west are 22 to 11

keeping with the different proportions of the various water masses contributing to the mean. The mean temperature rose 0.5°C and the mean salinity decreased 0.5‰ (Table 8).

Thus, the changes observed during the unusual occurrence of flow reversal in the strait appear to result from a selective damming of the system. Even with south net transport there is a region of northerly flow near Cape Prince of Wales, although much slower than normal (see Fig. 48b). Where southerly flow occurs, water that has transited the strait is returned into the strait region, and this water seems preferentially to consist of the two easternmost water masses. The total effect is to displace Anadyr Water with Alaskan Coastal and Bering Shelf waters, which have in turn interacted by greater layering and mixing than usual.

We note that there is no evidence of any water mass of a more northerly (Chukchi Sea) origin in Bering Strait during or after the time of south

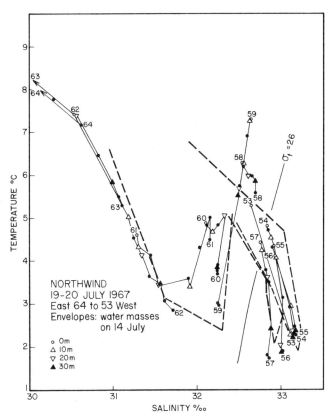

42. T-S diagram of Bering Strait section, *Northwind*, 19–20 July 1967: station numbers in order from east to west are 64 to 53

transport. However, the presence of such water in the northern Bering Sea has been postulated previously (Meilakh 1958), and evidence of cold, low salinity Siberian Coastal Water was observed close to Cape Dezhneva in 1933 (Ratmanov 1937a). The water was not penetrating the strait.

We feel that with large quantities of Bering Sea Water masses being transported north under normal conditions and dominating the southern Chukchi Sea (see Chap. 4), prolonged flow reversals would be required to flush this water south out of the system. We conclude that under present conditions, significant quantities of Chukchi Sea Water do not reach to the south of Bering Strait.

There seem to be no lasting effects of the flow reversal on the water masses and their properties; on resumption of normal flow, they resume their normal characteristics and distributions in the strait.

February 1968

We believe that the southerly flow observed by the *Northwind* in February 1968 in the Strait of Anadyr and the northern Gulf of Anadyr can also best be explained as an anomalous event.

Present evidence is that in winter as in summer, northward flow predominates in Bering and Anadyr straits. For example, this is the only way to explain the saline water present from Anadyr to Bering Strait throughout the year (cf. Figs. 30, 33), for it cannot come from east of St. Lawrence Island. The northerly flow is also local knowledge, e.g., the Eskimos on the eastern end of St. Lawrence Island relate the predominantly northward flow of the ice beyond about a mile from the shore, and depend on it in their walrus hunting. On the other hand, southerly ice movement in winter in Anadyr Strait has also been observed on an occasional basis, both by Eskimos and by persons involved in ice drift monitoring.

The current measurements at 30 m depth from the February 1968 *Northwind* cruise are shown in Fig. 44. Also included are six measurements made during the 1969 *Staten Island* cruise. Where measurements are from approximately the same location, just west of and to the southwest from St. Lawrence, they suggest a similar circulation pattern at the different times, with southwesterly flow through the eastern Strait of Anadyr. We note that current measurements from this area in summer all have shown north flow. The winter measurements thus appear to contradict a more or less continual supply of Anadyr Water into the Bering Strait region.

Some features of the hydrographic conditions in the Gulf of Anadyr at the time of the current measurements are summarized in Fig. 45 (cf. Fig. 19). Temperature-salinity correlations show that the Anadyr Water at that time lay in the salinity range 32.75 to 33.00°/$_{oo}$. As all water columns were

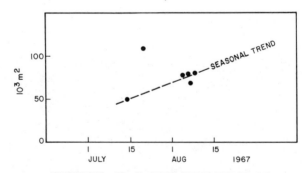

FRESHWATER AREA IN BERING STRAIT SECTION 1967

43. Fresh water area in Bering Strait section for the six sections occupied during summer 1967

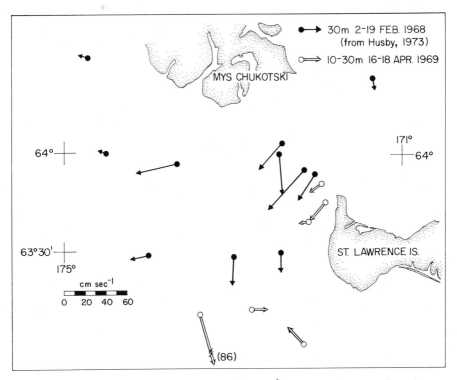

44. Current measurements near Strait of Anadyr at 30-m depth from *Northwind*, February 1968, and from 10 to 30-m depth from *Staten Island*, April 1969

quite homogeneous, typical of winter, Anadyr Water is distinctly separated from regions where the water was all <32.75‰ or all >33‰. The water >33‰ is the isolated high-salinity, and thus high-density, water of the northern gulf. In Fig. 45, this water is seen to occupy a larger area off Kresta Bay than in midsummer 1970 (cf. Figs. 14 and 15). Conversely, the area occupied by this water in September 1962 was even smaller (Fig. 19). We surmise from these data a possible seasonal cycle, the dense water being renewed in winter and then, over the summer, being progressively reduced in volume. Even though water mass analysis failed to show much interaction between Anadyr Water and this dense water in summer, there undoubtedly is some interaction, and it may be greatest in winter. In February 1968 the dense water occupied part of the water column at the station south of Cape Chukotski (Fig. 45), but was not present at adjacent stations, suggesting some sort of intermittent eastward flow.

The water <32.75‰ is the progenitor of the cold spot. Normally it lies more centrally in the gulf, separated from Siberia by Anadyr Water, but in

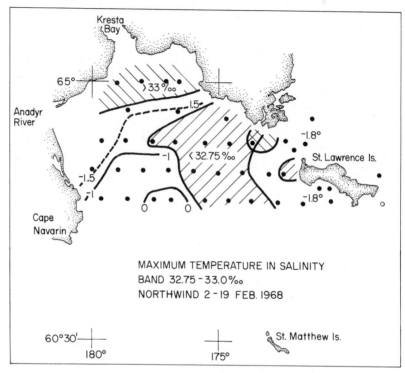

45. Hydrographic conditions in Gulf of Anadyr, 2–19 February 1968 (*Northwind*). Hatching isolates areas where all water in column was $S > 33°/_{oo}$ and $S < 32.75°/_{oo}$, and isotherms show maximum temperature in water columns with $32.75 < S < 33.0°/_{oo}$

February 1968 it had pushed north and effectively blocked any eastward flow of Anadyr Water. The latter, however, was entering the gulf from the south as usual, as indicated by the isotherms of maximum temperature.

The water northeast of the Strait of Anadyr was in the salinity range of Anadyr Water, and near freezing, i.e., at the temperature measured to be normal for this water mass in winter near Bering Strait. There appeared to be a continuity of this water south through the Strait of Anadyr and into the region south of St. Lawrence Island.

Our interpretation of these observations is therefore that they result from a temporary reversal of the predominant northward flow. This would have spread the water of the cold spot west and north from its normal position, as well as advected Anadyr Water from the region between St. Lawrence and Bering Strait (where it had been cooled to the freezing point), south through the Strait of Anadyr and into the region south of St. Lawrence. The experience of those familiar with the local ice conditions suggests that while the southerly flow in winter is anomalous, it is not an extreme rarity. Thus intermittent flow reversals could bring quantities of the higher salinity Anadyr Water into the region south of St. Lawrence, and thus contribute to the ultimate production of Bering Shelf Water.

A word of caution is necessary in regard to interpreting water motion from observations of ice motion. Wind on occasion can move ice in directions contrary to those of the surface water layer, and likewise the surface water layer can move contrary to the bulk transport (cf. Bering Strait sections, Fig. 48).

3

CURRENTS

GENERAL FLOW FIELD

The more detailed flow regime throughout the region can be described from those regional quasi-synoptic surveys that included current measurements. Three surveys covered large portions of the region: 1960 *Brown Bear*, the eastern side (east of 170°W) from St. Lawrence Island to nearly 71°N; 1968 *Staten Island*, east-west sections across the sea (to ~22 km off Siberia) from St. Lawrence to north of Bering Strait; and 1972 *Oshoro Maru*, sections across Bering Strait, and in the Chukchi Sea from Cape Lisburne to Siberia, and from west of Herald Island to Pt. Barrow. At all stations the vessels were anchored and current measurements were made with a deck readout current meter; the station time was <1 hour. Figure 46 presents the 1960 current vectors at 5 m and 20 m; and Fig. 47 presents averages of the 0 to 10 m, 10 to 30 m, and 30-bottom layers for 1968. (For *Oshoro Maru* results see Chap. 4.)

In elucidating the general features of the flow field from these data, due regard must be given to possible time variations in currents. Fleming and Heggarty reported that, based on results from the 1959 *Brown Bear* cruise, tidal currents were not significant over much of the area and that currents at any individual station were relatively steady in both speed and direction. We have earlier (1966) reported that within Bering Strait short-term variations in direction are not large, but that speed fluctuations are as much as 50% of the mean speed, with an important periodicity of 12 to 13 hrs. A four-day current record from north of Bering Strait in March 1968 showed an important fluctuation due to the diurnal tidal wave, but there was also significant variation with a time scale >2 days (30% of the mean speed) (Coachman and Tripp 1970). Some stations have shown large directional variation, particularly in the upper layer; e.g., station 34, 8 to 9 August 1933 (off Cape Dezhneva, Ratmanov 1937a), showed 190° variation during one day; and in Figs. 38a, b we see directional changes >90° in a few hours.

We conclude that there may be significant local velocity variations, and

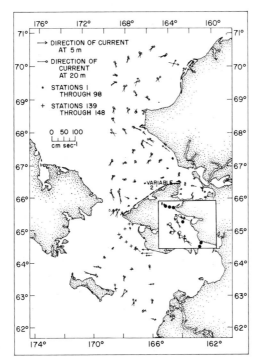

46. Currents measured at 5 m and 20 m from *Brown Bear*, 26 July–28 August 1960 (from Fleming and Heggarty 1966)

these are discussed in more detail in the next section. Because of this, when describing the mean flow field, current measurements from individual stations cannot be treated independently, but must be considered together with values from adjacent stations. As a large proportion of the variations appear to have time scales of 12 hours or less, transport calculations based on sections requiring a day or longer to complete may not be too seriously in error.

The major feature of the mean flow is of course the general north direction, but the flow speeds are not uniform across or along the system. In the straits (Bering, Anadyr, and between St. Lawrence Island and Alaska), flows are almost always swifter than away from these bathymetrically constricted areas. Also, as noted by Fleming and Heggarty, flows along the south sides of the major westward-projecting promontories of the Seward Peninsula and the Pt. Hope-Cape Lisburne peninsula are markedly swifter than the currents farther west. Thus, as the northward flow converges under the constraints of the straits and the westward-projecting land masses, it is accelerated.

Fleming and Heggarty supposed there would be a pattern to these accelerated flows in the straits, that is, that the greater speeds would be evidenced in the more easterly portions of the straits. Our more extensive data show this to be true only in Bering Strait; there the swiftest flows are always observed in the eastern channel west from Cape Prince of Wales. In the other straits this does not seem to be true. In the Strait of Anadyr the swiftest flow appeared in midchannel both in July 1968 (Fig. 47) and during the case of south flow in February 1968 (Fig. 44). In the strait east of St. Lawrence the fastest currents were in the eastern part of the section in July 1968 (Fig. 47) but in the western part in August 1960 (Fig. 46).

Figure 48 shows isotachs of the flow for all available detailed current cross-sections in Bering Strait. A variety of conditions are encompassed in the eleven sections, including a range of transports from >2 Sv north to 0.2 Sv south. There are several cases of local flow reversal in the upper layer, but in all sections the swiftest flows are west of Cape Prince of Wales. Thus, the pattern of horizontal shear in Bering Strait seems to be relatively invariant. (A velocity minimum just west of Fairway Rock also seems to be a frequent feature of these sections.)

In the regions away from the straits and the two major promontories, the currents are markedly slower and more variable in direction. Some of the features may be semipermanent eddies or meanders in the flow field, though because of time-dependent variations these cannot on the whole be rigorously defined by the present data. Our data do tend to confirm the presence of one major meander and two semipermanent eddies inferred by Fleming and Heggarty. As these lie north of Bering Strait, they are discussed in Chap. 4.

The more open central region north of St. Lawrence Island has been suggested as the site of a semipermanent eddy or gyre (Barnes and Thompson 1938). However, examination of all data, both currents and water mass distributions, gives no indication of any steady gyrelike circulation in the region. The currents are less in magnitude and more variable in direction than elsewhere. The continuity of water mass properties suggests that the water in summer comes from around both ends of St. Lawrence and shows modification consistent with general northerly flow.

FLUCTUATIONS IN THE CURRENT FIELD

We now need to examine the current field for its temporal variability, particularly in the tidal frequency bands. The fact that the tidal range in much of the region is small, e.g., a fraction of a meter in Bering Strait, and probably about a meter in Long Strait, does not *a priori* mean that the

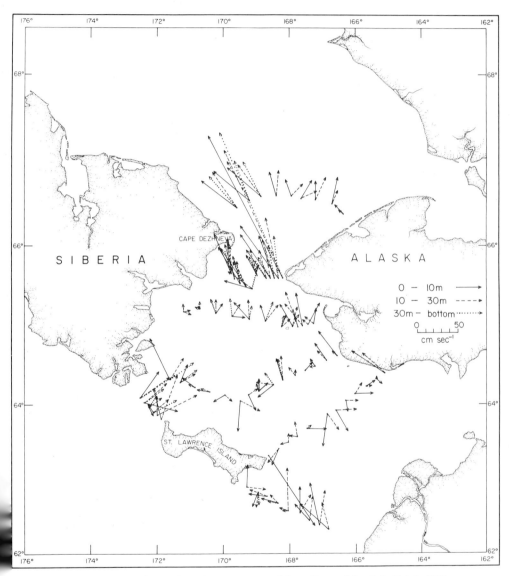

47. Currents averaged over three layers measured from *Staten Island*, 9–19 July 1968

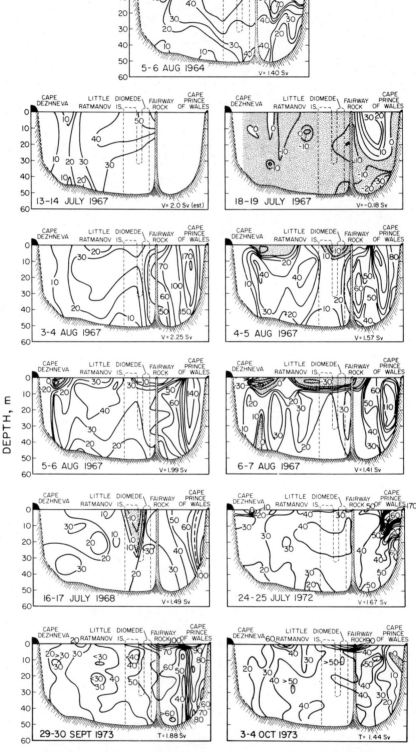

48. Isotachs of north-south flow for the eleven detailed Bering Strait sections

tidal currents are negligibly small. (Where locally amplified by the basin geometry, tidal currents might in fact be expected to be quite large, e.g., Kresta Bay.) The semidiurnal frequency band may be somewhat complicated by the fact that one can, in general, expect it to be contaminated by inertial motion. At the latitude of Cape Navarin the inertial period is 13.6 hrs, in Bering Strait 13.2 hrs, and in Long Strait 12.8 hrs.

Prior to 1964 work on periodic current variations, primarily in Bering Strait, is inconclusive (cf. Coachman and Aagaard 1966, for a historical review). However, in August 1964 current measurements made in Bering Strait over nearly two days from the *Northwind* did show a semidiurnal variation with amplitude of order 10 cm sec^{-1} (Coachman and Aagaard 1966). The measurements were not suitable for a meaningful periodic analysis.

The 1967 Measurements

Much better current records for these purposes were made in July 1967 when two moored arrays of two current meters each were deployed for about 17 days in the vicinity of Bering Strait, one approximately 100 km north and the other 90 km south. The meters were Braincon model 316 which photographically recorded the integrated speed and directional variation over a 9-minute interval. Tow tank calibrations and meter intercomparisons indicate a speed accuracy of about ±1 to 2 cm sec^{-1} at speeds less than 10 cm sec^{-1} and ±10% at higher speeds. The current meters have a calibration threshold of 2.5 to 3.0 cm sec^{-1}. The directional accuracy is about ±15 degrees of arc relative to the internal magnetic azimuth. The circumstances of the records are summarized in Table 9.

The vectorially averaged currents are shown in Fig. 49. The mean flow was northerly at all meters, with a slight easterly tendency south of Bering Strait and westerly north of the strait. At the northern location, the mean current was equally strong at the two depths and differed in direction by only 10 degrees of arc. At the southern location, the directional difference was about the same (9 degrees and in the opposite sense), but

TABLE 9
Moored Current Meters

Meter No.	Latitude	Longitude	Water depth m	Meter Depth m	Start of Record GMT	Termination of Record GMT
156	66°36'N	168°19'W	32	13	0600 11 July 1967	1430 28 July 1967
157	66°36'N	168°19'W	32	25	0600 11 July 1967	1430 28 July 1967
158	64°55'N	168°40'W	48	13	0415 12 July 1967	0025 29 July 1967
159	64°55'N	168°40'W	48	35	0415 12 July 1967	1200* 20 July 1967

*Estimated.

the speed at 35 m was over 12 cm sec^{-1} less than at 13 m. Part of this difference might be associated with the difference in observational span (cf. Table 9); i.e., the observations at the lower meter covered only about the first half of the time span of the upper meter, and the upper meter showed higher speeds during the second portion of its observational span. Examination of the trend at the upper meter indicated that this could account for only about one-third of the observed mean speed difference, however. The conclusion must therefore be that the only significant mean vertical shear was at the southerly location, where its value was about 3 to 4 × 10^{-3} sec^{-1}. This shear is probably larger (perhaps by about one-half order of magnitude) than can be accounted for by the mean geostrophic shear. Presumably much of the observed shear is associated with baroclinicity in the direction of flow.

We note that the mean flows might be considered fairly typical for their respective locations, being northerly and of magnitude 20 to 30 cm sec^{-1}. There is therefore no reason to think that any temporal variations observed in the current meter records are atypical. Before further analysis,

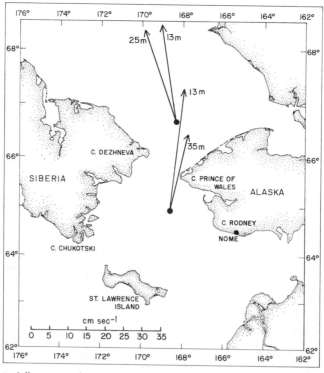

49. Vectorially averaged currents at the moored arrays, 11–29 July 1967

the current records were vectorially averaged over one-hour intervals to suppress noise and facilitate data processing.

Flow Reversals

Even though the mean flow was northerly everywhere, the record at each meter shows times of southerly flow. These events are listed in Table 10. Other than that the southerly flow occurred during the same general period (21 to 25 July) at three of the four meters, there is very little similarity between the events. Even at the same mooring, the instances of southerly flow do not in general appear to be vertically coherent. For example, during the period of pronounced southerly flow at meter 159 (16 to 18 July), there was not even a decreasing trend in the northerly component at meter 158, located only 22 m higher on the same mooring. Taking the records as a whole, at meter 158 southerly flow occurred only 1 percent of the time, but nearly 10 percent at meter 159 despite that meter not operating at the time no. 158 showed southerly flow. Similarly, while the maximum southerly velocity component at meter 158 was only about one-sixth as great as the mean meridional component (directed northerly), it was larger than the mean component at meter 159.

We note also that at the northern mooring the southerly flow occurred as part of an anticyclonic rotation of the current vector. However, at the southern mooring the southerly flow component was phased to make the vector pass through an arc segment with an alternating sense of rotation. Figure 50 shows two representative hodographs.

To provide a dynamic explanation for the southerly flow events is problematic. They are not related in any obvious way to the local wind regime, for during the period of strongest northerly winds (~30 knots on 19 July), the flow was northerly everywhere; conversely, during a general period of southerly flow (21 to 23 July) the winds were variable and relatively weak.

Probably the greatest obstacle to a reasonably simple explanation of the southerly flow is its extraordinary lack of coherence over very small vertical distances. This means not only that the flow reversal does not represent a general change in the whole system, but also that it cannot be caused by a more localized barotropic disturbance.

We can visualize only one mechanism as possibly responsible for the southerly flow events, namely quite thin baroclinic eddies. If these were quasi-geostrophic, they would, in order to account for the observed shears, require isopycnal slopes as high as 3 to 4×10^{-2} for vertical density differences of 2×10^{-4} gm cm^{-3} (0.2 in σ_t). We note that modeling a specific reversal event in terms of eddy motion requires something more complex than a single eddy being advected past the current meter at a

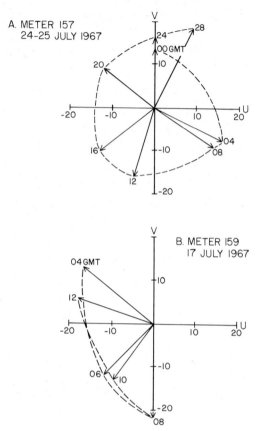

50. Representative hodographs during flow reversals at the moored arrays: speeds in cm sec^{-1}

constant rate. Rather, in order to account for the observed vector rotation, the eddies must occur in multiples, with adjacent eddies having opposite senses of rotation (i.e., a vortex street) and/or the rate of advection of the eddies past the meter has to be quite variable. We feel that there are too many uncertainties in such reconstruction to justify size estimates of the eddies that might be involved.

Linear Trends

In similarity with the vertical incoherence of southerly flow events, the long-term trends in the records from different meters on the same mooring also appear to be curiously independent, as shown in Table 11. Thus the northerly set at meter 156 shows a slowing tendency, while at meter 157, a quickening tendency is apparent. In fact, only the easterly velocity

TABLE 10
Records of Southerly Flow

Meter No.	Date July	Duration, Hours	Maximum Southerly Component (hourly mean), cm sec^{-1}
156	21	3	11.2
156	22	11	17.8
156	22–23	6	11.6
156	23–24	3	5.5
157	22	2	1.6
157	24	13	18.0
157	24	1	4.5
158	23	2	5.4
158	25	2	0.8
159	16	2	12.4
159	17	6	21.6
159	17	9	16.5
159	18	2	4.3

components at the southerly mooring have even the same sign in the trends. If the trends of Table 11 reflect trends in the pressure gradient force which drives the system, it implies that the baroclinic and barotropic contributions to the pressure gradient act rather independently.

Semidiurnal Variations

While the 1967 current records are of insufficient length to provide really good and statistically reliable resolution through spectral techniques, the spectra prove both suggestive and interesting. Typically the power spectra give a frequency separation of 0.17 cycle per day for 15 degrees of freedom; this separation corresponds to about one hour in

TABLE 11
Trends in the Current Records from the Moored Meters

Current Meter No.	Velocity Component	Linear Trend cm sec^{-1} hr^{-1} × 10^3	Percent Change per Day
156	north	−1.07	−2.6
	east	−0.75	−1.8
157	north	0.24	0.6
	east	0.25	0.6
158	north	0.68	1.6
	east	−0.39	−0.9
159*	north	−2.79	−6.7
	east	−0.47	−1.1

*Current record approximately one-half as long as at the other meters.

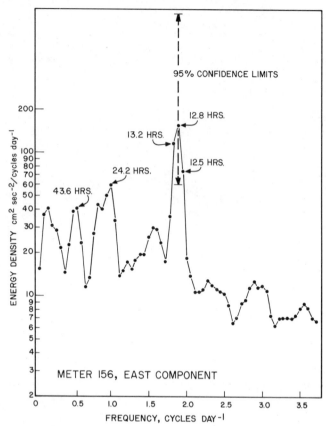

51a. Power spectrum for current record at meter 156, east component

the semidiurnal band. Since we are using the spectra in a suggestive sense, we have chosen to increase the frequency resolution as much as feasible, at the expense of decreasing the degrees of freedom to about 5. This gives 95% confidence limits for the power spectra of only 0.39 to 6.02 and for the coherence spectra of 0.73, but a frequency separation corresponding to 20 minutes in the semidiurnal band.

In general, the spectra vary in detail, but all have semidiurnal peaks. Figures 51a, b show the power spectra for the east and north components at meter 156. The various possible semidiurnal frequencies are clearly not separated by the analysis, but the suggestion is that the inertial frequency (corresponding to 13.08 hours) may be of primary importance, as distinguished from, say, the M_2 constituent (12.42 hours). This is true for both components. We note that the spectral estimates indicate the energy density of the east component to be about three times as large as that of

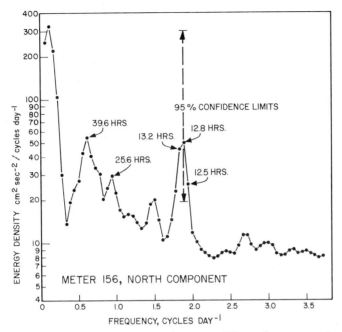

51b. Power spectrum for current record at meter 156, north component

the north component so that the ratio of the velocity components would be about $\sqrt{3}$. The coherence spectrum has a peak at 12.8 hours of form similar to the power spectra with a coherence of 0.87, which is well above the 95% confidence limit; the north component leads in phase by 125°. On the basis of the spectra only, we might therefore venture to predict a semidiurnal current ellipse rotating clockwise with major axes oriented WNW/ESE, with a maximum speed of about 18 cm sec^{-1} and a major to minor axis ratio of about 2.5.

Figure 52 shows the velocity components of the semidiurnal period after processing by a numerical band-pass filter. The filter causes very little phase error, but it passes, 95 percent unchanged, a frequency band corresponding to about two hours in period. The major feature of the records is a pronounced semidiurnal oscillation, with the east component having an average amplitude of about 10 cm sec^{-1} and the north component about one-half as great; the north component leads by perhaps 130° on the average. The filtered records obviously contain more than a single harmonic constituent, since the phase and period vary somewhat, and the amplitudes are variously modulated. However, the mean period is about 12.8 hours, which corresponds identically to the spectral peak. The current ellipses constructed from each cycle of the

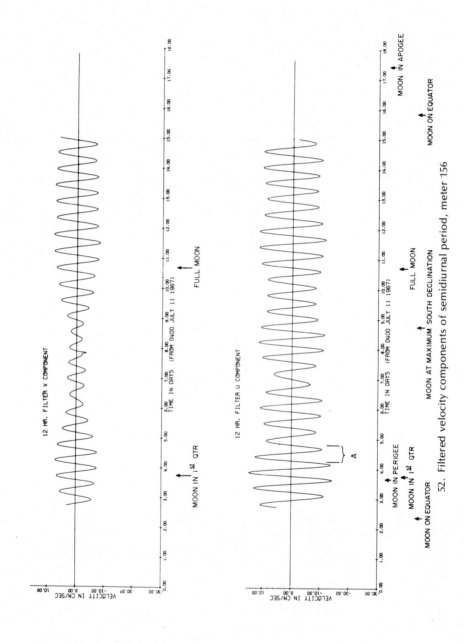

52. Filtered velocity components of semidiurnal period, meter 156

filtered records are essentially similar, showing clockwise rotation and an ellipse elongated in a direction between SE/NW and E/W. An example is seen in Fig. 53, constructed from the fourth cycle in Fig. 52 (marked A). Except for a lower amplitude, the ellipse corresponds rather closely to that predicted from the spectra.

At meter 157, the spectra show the north component of velocity to be the stronger, with a peak at 12.7 hours, while the east component has several peaks in the semidiurnal band, including one at the inertial frequency. The component coherence also has a peak at the inertial frequency, but is just under the 95% confidence limit. On the other hand, the phase spectrum indicates counterclockwise rotation, rather than the clockwise rotation of inertial motion. The filtered data show an average period slightly greater than the inertial, and a changing phase relationship corresponding to current ellipses rotating in both senses. The highest amplitudes (~8 cm sec^{-1}) occurred during a two-day interval centered on 22 July, just after full moon, with the components nearly in phase, about equal in amplitude, and of period 13 hours. This would correspond

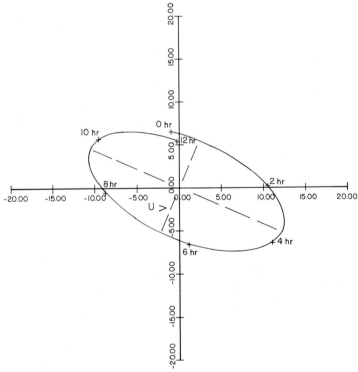

53. Current hodograph from fourth cycle in Fig. 52: speeds in cm sec^{-1}

to a nearly alternating semidiurnal current directed NE/SW with a maximum speed of 11 to 12 cm sec^{-1}.

The coherence between meters 156 and 157 in the semidiurnal band is quite good, being 0.81 (at 12.7 hours) for the east component and 0.76 (at 13.1 hours) for the north component, both well above the 95% confidence limit.

At the southern mooring, there were also prominent semidiurnal oscillations. Meter 158 registered velocities with a spectral peak at 12.4 hours in the east component and 12.0 hours in the north component. The filtered records show a semidiurnal oscillation with the phase varying such that the current ellipse rotated in both senses, and modulated amplitude ranging from about 1 to 12 cm sec^{-1}. The modulation is similar to the one at meter 156, with a maximum near the time of lunar quadrature and another maximum about a week later. The current ellipse was in the majority of instances oriented between NE/SW and N/S, and frequently degenerated into nearly rectilinear motion. The component coherence was only 0.68, further strengthening the impression of a rather complicated semidiurnal band.

In passing, we note the somewhat curious fact that the north components of velocity at meters 156 and 158 were coherent at the 95% confidence level for a period of 12.7 hours. If one uses the phase spectral estimate, which has meter 156 leading 158 by 54° ~1.9 hr, this corresponds to the semidiurnal tidal wave progressing southward through Bering Strait with celerity approximately appropriate to the depth. However, examination of the original and filtered records shows a variable phase relationship between the north component at the two meters, so that the issue of the tidal wave direction in Bering Strait must remain unresolved.

The record at meter 159 was only about one-half as long. The spectra show a well-defined semidiurnal peak only for the east component, at 12.4 hours. However, the filtered records have a semidiurnal oscillation in the north component also, of from 2 to 5 cm sec^{-1} as contrasted to 5 to 10 cm sec^{-1} in the east component, the phase difference varying around 180°. The current ellipse is therefore generally quite elongated and oriented NW/SE; the less degenerate ellipses rotate counterclockwise.

We are thus left with the clear indication that both north and south of Bering Strait, there are prominent semidiurnal oscillations in the current field. The net (peak-to-peak) amplitude of these oscillations is typically about 10 cm sec^{-1}. The current hodographs show varying eccentricity (in some cases degenerating into rectilinear motion) and orientation, and both senses of rotation, suggesting the interplay of several different semidiurnal constituents. In general, we must probably look to tidal explanations of these oscillations, for their characteristics are on the whole inappropriate to inertial motion. This is not to say that there are no inertial oscillations in the records, but rather that such oscillations are not separable in the present data. On the other hand, no simple tidal model will suffice either.

For example, Fig. 54 shows the modulations of the filtered semidiurnal current oscillations. These were computed by taking sums of the daily mean oscillation range for each component; the results from meters 156 and 157 were averaged to represent the northern mooring, and the results from meter 158 were averaged over two days to represent the southern mooring (the record at meter 159 was too short to be utilized effectively). This procedure smooths the data without significant loss of information about the modulation of the major semidiurnal constituents. The form of the modulation is rather similar at the two moorings, with a minimum on 18 July and a maximum four days later. Only initially, however, are the oscillation ranges comparable, that at the southern mooring subsequently being considerably less (so that the trends of the two curves are in fact opposite, increasing at the northern mooring and decreasing at the southern).

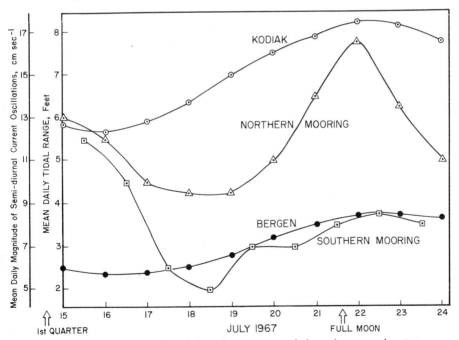

54. Modulations of filtered semidiurnal current records from the moored meters, and daily tidal range at Kodiak and Bergen

A priori, we would expect Bering Strait to be a region of transition between influence of the Atlantic and Pacific tidal waves. The Arctic Ocean waves are thought to come from the Atlantic, principally through the Greenland-Spitsbergen passage, and have been identified at least as far as the central Chukchi Sea (Sverdrup 1927). We have thus plotted in Fig. 54 the mean daily tidal range at Bergen, Norway as well as that at Kodiak; the latter is taken to represent the Bering Sea waves (Kodiak is the tidal prediction base for a number of stations on St. Lawrence Island and in Norton Sound). The modulation of the range is similar for the two stations, with a minimum about 16 July and a maximum six days later, just after full moon. The tidal current curve (mean daily peak-to-peak amplitude) for Unimak Pass, which is the prediction base for tidal currents in the region Sledge Island to Fairway Rock, shows a form nearly identical to the Kodiak tidal curve. However, the modulation of the mooring records resembles none of these tidal curves, except for the maximum near 22 July. Thus, just as with the variety of semidiurnal current hodograph characteristics, the modulation does not appear to lend itself to a simple modeling.

In the northern Bering Sea, the tides themselves appear rather complex. In Norton Sound the tide is chiefly diurnal, with a cyclonic progression around an amphidromic point. This is clear from the tide tables and also from published cotidal charts (e.g., Office of Climatology and Oceanographic Analysis Division, 1961). Westward from Norton Sound, the tide includes semidiurnal components, so that the tide tables specify a mixed tide at St. Lawrence Island (reference stations are Nushagak and Kodiak), while on the Siberian shore the reference station for the Gulf of Anadyr and northward is Pusan, which has a uniformly semidiurnal tide. Other than in Norton Sound, the charts of cotidal lines referred to above do not agree with the distributions indicated by the tide tables. For example, the chart shows the cotidal line as being parallel to the south shore of St. Lawrence Island, while the tide tables indicate a time difference from east to west of 7 to 8 hours. The tables thus suggest a cyclonic amphidromic system immediately south of St. Lawrence. Between St. Lawrence and the Alaskan mainland the cotidal line trends WNW/ESE, corresponding to a progressive wave entering the northern Bering Sea through that passage. The tables could also be interpreted as indicating a cyclonic amphidromic system in the Gulf of Anadyr. Northward from St. Lawrence Island, the semidiurnal tide seems to be increasingly predominant. Tide records made by the United States Coast and Geodetic Survey (U.S.C.&G.S., unpublished data) in summer 1968 show tides at Nome rather closely resembling the Dutch Harbor tides: semidiurnal for some time subsequent to the moon passing the equator, but diurnal near the time of maximum lunar declination. Therefore, the choice of Dutch Harbor as the reference station for Nome is reasonable. However, at Pt. Spencer, some 125 km NW of Nome, the tide is largely semidiurnal (U.S.C.&G.S., unpublished data), closely resembling the Nome tide when the latter is semidiurnal and lagging behind it by 3 to 5 hours, but taking on only a slightly mixed character when the Nome tide is diurnal and being quite different in phase. We note that Bloom (1956) has stated the tides at Wales to be primarily semidiurnal.

In the Chukchi Sea the tides have traditionally been interpreted as progressive waves coming in from the deep Arctic Basin, and there is nothing in the more recent tidal height data to contradict this view. The tidal reference station for the Siberian coast north of Bering Strait is Pusan (semidiurnal tide) and for the Alaskan coast Kodiak (mixed tide). This agrees with the Arctic tide type chart (U.S. Navy Hydrographic Office 1958) which shows semidiurnal tides along the Siberian coast, but mixed tides at high lunar declination for the Alaskan coast and the western Canadian archipelago. We note, however, that the cotidal chart appears to be in error in the Chukchi Sea, where it indicates a tidal wave celerity much slower than that found by Sverdrup (1926) or indicated by the current observations of Coachman and Tripp* (1970). Nor does the cotidal chart agree with the tide tables, showing, for example, a high water difference between Cape Billings and Cape Schmidt of over an hour, whereas the tide table gives 5 minutes. From the tidal height observations, we would therefore expect that semidiurnal current oscillations would predominate over diurnal in the Bering Strait region, and that several different tide waves are present, producing rotary currents. Other than in this very general way, the extremely complex oscillations observed at the anchored meters are not readily interpretable in tidal terms.

Variations Shorter Than the Semidiurnal

The power spectra for meter 159 show a very sharp peak at 6.2 hours, in the east component nearly as large as the semidiurnal and in the north component over twice as large. This oscillation is extremely clear and regular in the first few days of the unfiltered record, particularly in the north component, where it has an amplitude of about 4 cm sec^{-1}; the components are phased about 140°. Later in the record it becomes less

---

*Those measurements, made 22 to 24 March 1968, showed diurnal oscillations of 5 to 6 cm sec^{-1} in the southeastern Chukchi Sea, which is comparable to those observed at meter 156 (see below). However, the semidiurnal oscillation was less than 1 cm sec^{-1}. Maximum southerly lunar declination was during the afternoon of 21 March, and it may thus well be that the relatively large diurnal component was associated with a marked diurnal inequality in the tides at that time.

regular both in period and phase, but its amplitude increases to as much as 15 cm sec^{-1}. The coherence spectrum of the components also has a pronounced extremum at 6.2 hours, reaching 0.67. The sharpness of each spectral peak and its exact correspondence to the first harmonic (M_4) of the lunar semidiurnal frequency are by themselves suggestive of a shallow-water tidal constituent. However, a number of factors speak against this explanation, among them: (1) the water depth is too great to expect significant shallow-water effects; (2) the quarter-diurnal amplitude on the whole is as large as, or at times even larger than, the semidiurnal amplitude; and (3) there is no evidence of persistent quarter-diurnal oscillations at meter 158 located only 22 m higher on the mooring.

Instead we suggest that the oscillations result from the advection of baroclinic eddies past the current meter, much as discussed for the case of southerly flow, but with the difference that either these eddies are weaker than those associated with the southerly flow, or the advection rate of the eddies is much greater, or both. In any case, a current hodograph such as shown in Fig. 55, with a series of cyclonic loops, can in its main features be rather readily reconstructed from a sequence of eddies

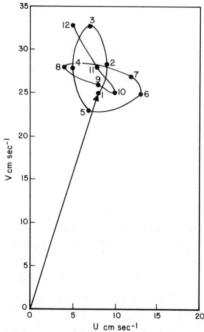

55. Current hodograph for selected portion of current record at meter 159, showing quarter-diurnal variations

being advected with the mean flow. The eddies of Fig. 55 would have a radius on the order of 2 km. We remark on the great frequency of occurrence of the phenomenon, and note in this connection that meter 158 also recorded large oscillations with periods shorter than the semidiurnal. These were mostly longer than the M_4 period, irregular but of very large amplitude. For example, the power spectrum for the east component at meter 158 has a peak at 9.3 hours nearly as large as the spectral amplitude of the semidiurnal oscillation.

The observations are not conducive to illuminating the problem further, but the matter of baroclinic eddies in this region certainly merits an appropriately designed observational program.

Diurnal Variations

Diurnal variations in the current records can be seen for all current meters, but they are in each case considerably weaker than the semidiurnal oscillations. Examination of the spectral estimates and filtered records leads to the following conclusions.

At meter 156 the diurnal oscillations were quite clear and the components varied between 3 and 11 cm sec^{-1} in amplitude. The record is of insufficient length to define the modulation satisfactorily, but the maximum amplitude appears to have occurred about halfway between first quarter and full moon (which was the time of minimum semidiurnal amplitude), and the minimum diurnal amplitude was about two days after full moon. The diurnal current vector rotated clockwise and the ellipse was oriented NE/SW.

At meter 157 the diurnal oscillations were weaker, with amplitudes of 1 to 2 cm sec^{-1}. Only the north component was modulated, the modulation being approximately opposite in phase to that at meter 156. The ellipse was quite elongated and oriented NW/SE, and the current vector rotated in both directions.

Oscillations with somewhat similar characteristics were found at meter 158. The component amplitude, which varied between 1 and 5 cm sec^{-1}, was modulated largely in V and with the same modulation phasing as at meter 157, and the current ellipse was oriented WNW/ESE. However, in the majority of instances the current vector rotated counterclockwise.

At meter 159 the east component of the diurnal oscillation was about 5 cm sec^{-1}, while the north component was 1 to 4 cm sec^{-1}. The current vector rotated counterclockwise and the ellipse was oriented NE/SW.

Longer–Period Variations

The current records also show variations of periods longer than the diurnal. These are not necessarily of a persistent harmonic nature, but may typically show rather regular fluctuations persisting over several cycles; or to be of sufficiently long time scale to go through a cycle or less but have a regular and symmetric form suggestive of harmonic oscillations. For example, the variation of the semidiurnal current oscillations in Long Strait discussed by Coachman and Rankin (1968), which was associated with the fortnightly tidal cycle, is representative of the latter category.

The spectra from meters 156 to 158 contain a maximum in the frequency band corresponding to an average period of about 42 hours. Variations of this time scale are apparent in the original records, and numerical filtering brings them further into prominence. Meter 156 shows a series of rather regular oscillations of amplitude 5 to 10 cm sec^{-1} and a period of about 42 hours. A number of cycles have components nearly in phase, corresponding to reciprocating motion, whereas at meters 157 and 158 the east component lags. Meters 157 and 158 also show a series of oscillations of shorter period (30 to 36 hours) and variable amplitude (2 to 13 cm sec^{-1}), followed on 23 to 24 July by a large pulse of speed amplitude 27 cm sec^{-1}. Beginning 25 July there was a second large pulse of 2 to 3 days duration. These events are reflected in a 0.88 coherence maximum between the north components of meters 157 and 158, with a period of about two days and meter 157 leading in phase. A reasonable estimate of the phase difference between the two meters is 10 to 20 hours for the large pulses, which for a meter separation of 190 km corresponds to a pulse propagation rate of 2.6 to 5.3 m sec^{-1}. This contrasts with a free shallow-water wave propagation rate of about 20 m sec^{-1} (for the internal shallow-water wave the rate would be less than 0.5 m sec^{-1}). The implication is that the pulse is not a freely traveling disturbance. On the other hand, the calculated propagation rate is considerably faster than the mean flow, so that it is not likely to be the result of the same feature advected past the two meters by the mean flow. The possibility remains of a forced pulse, but traveling atmospheric disturbances that might cause such a pulse usually move at rates closer to 10 m sec^{-1} or so. For example, an atmospheric front across the northern Bering Sea that had been stationary for three and one-half days, began moving northward late on 24 July and moved through Bering Strait early on 25 July, covering about 260 km in 6 hours, corresponding to a travel speed of 12 m sec^{-1}. Therefore, no simple model of a forced disturbance appears very appealing.

There is also the question of the origin of the slightly longer-period fluctuations, with a typical time scale of one and one-half days. These were particularly regular at meter 156 (42-hour period). One might inquire into this possibly being a natural oscillation period for the system.

However, the fundamental seiche mode of this period is appropriate to a system on the order of 1500 km, which is the scale of the combined meridional extent of the Bering-Chukchi shelf. Alternatively one might look to atmospheric excitation of the oscillations. For example, consider the winds at Wales during 11 to 30 July 1967 (Fig. 56). There is obviously an oscillation in the wind, particularly in the north component, with major wind reversals occurring within 1 to 2 days. These oscillations are not readily recognizable on the synoptic weather charts, so that their regional extent remains uncertain:* nonetheless, the suggestion is certainly that there are at least local atmospheric disturbances of time scale comparable to the current oscillations. It proves difficult to pursue the matter much further on the basis of the present observations.

The 1968 Measurements

In July 1968 three anchored current stations were made from the *Staten Island* in the northern Bering Sea, two of 30 hours duration [at 63°18'N, 168°23'W, east of Northeast Cape, St. Lawrence Island (2XX in Fig. 3), and at 63°59'N, 172°05'W, in the Strait of Anadyr (3XX in Fig. 3)] and one of 25 hours duration [at 65°12'N, 170°19'W, about 55 km SE of Cape Krigugan

*Smaller, but in several respects somewhat similar, wind variations can be seen in the Cape Lisburne records from the same period.

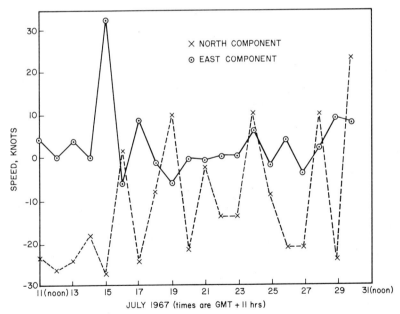

56. Wind speed components at Wales, 11–30 July 1967

(1XX in Fig. 3)]. At each station continual current profiles were obtained by lowering and raising a deck-read-out meter.

At the Northeast Cape station, taken 6 to 7 July, the mean current was in the range 35 to 40 cm sec^{-1} directed between NNW and N. In general, the speeds tended to be slightly greater in the upper 10 m than farther down. A semidiurnal oscillation is quite apparent in the record, typically being 10 to 25 cm sec^{-1} in amplitude in the north component and somewhat less in the east component. Particularly noticeable is the reduction with depth of the oscillation amplitude: in the layer 0 to 10 m it is over twice as great as in the layer 10 to 30 m. This effect is undoubtedly a result of density stratification, and examination of hydrographic casts made during this anchor station shows a mean density difference between the two layers of 0.39 in σ_t (see below). The east component lags by about 4 hours, so that the semidiurnal current vector is rotating clockwise. The record also appears to show some slight diurnal inequality. In this connection we note that maximum southerly lunar declination was early on 9 July; full moon occurred on 10 July.

At the Anadyr Strait station (shown in Fig. 57a, b), taken 8 to 9 July, the semidiurnal oscillations were extremely pronounced. They were superposed on a mean flow of about 55 cm sec^{-1} directed NE. At this location also, the amplitude of the oscillation decreased with depth; this was particularly noticeable in the north component, where the amplitude in the 0 to 10 m depth range was typically over 50 cm sec^{-1}, but less than 30 cm sec^{-1} in the 10 to 30 m range. The phase difference, if any, is not large, corresponding to a progressive wave passing through the strait. Conceivably, the record could be construed to show some diurnal inequality, but the data are ambiguous in this respect.

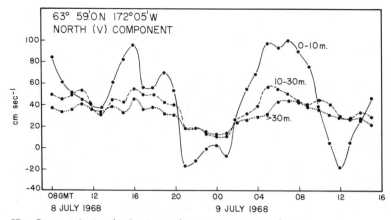

57a. Currents in Anadyr Strait, north component, 8–9 July 1968

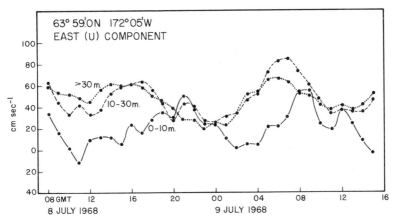

57b. Currents in Anadyr Strait, east component, 8–9 July 1968

A rather different situation prevailed at the northern station, taken 4 to 5 July. The mean flow was somewhat less, about 20 cm sec^{-1} toward N/NNE, but more importantly, any semidiurnal oscillations were sufficiently small not to be apparent in the current record, except perhaps in the upper (0 to 10 m) layer, where their amplitude, if present, was not over about 5 cm sec^{-1}. It would thus appear that the tidal flow is somewhat amplified through the two straits flanking St. Lawrence Island and subsequently reduced in the more open shelf area to the north.

The matter of reduction of tidal current amplitude with depth needs some further comment. At both the Anadyr Strait and the Northeast Cape stations, the considerable current decrease occurred in the presence of stratification. In Fig. 58 the shear in the semidiurnally oscillating current has been plotted as a function of density stratification. It is clear that the shear appears correlated with the degree of stratification, although the number of data is admittedly very small. The implication is that the internal tide is acting to reduce the effect of the external tide. If so, we should see oscillations of similar period in the depth of the isopycnal surfaces. That something like this occurs is indicated by Fig. 59, showing the depth of the isopycnal surface $\sigma_t = 26.0$ during the anchored station in Anadyr Strait. The approximate times of maximum and minimum current below 10 m are also indicated (compare Fig. 57a, b). The minimum in speed coincides approximately with the maximum elevation of the isopycnal surface and the maximum in speed approximately with the maximum depression of the isopycnal surface. If Fig. 58 therefore reasonably represents the effect on the tidal speed amplitude of stratification, then the ordinate intercept of 5×10^{-3} sec^{-1} should represent the frictional effect. We note that in the upper one-half of the water column,

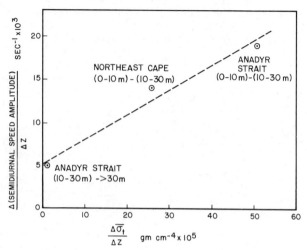

58. Shear in semidiurnal current oscillations as a function of density stratification, Anadyr Strait and Northeast Cape stations, 6–9 July 1968

the stratification effect is typically 2 to 3 times as great as the frictional effect, and only for relatively low stratification, such as at the Cape Krigugan station (1XX Fig. 3), are they approximately equal. If we assume a frictional stress on the order of 1 dyne cm^{-2}, then the eddy viscosity must be about 200 cm^2 sec^{-1}. This is in the range of values calculated by Fjeldstad (1936) for the tidal currents on the North Siberian shelf, namely 10 to 400 cm^2 sec^{-1}.

We note that at none of the anchored current meters (156 to 159) in the vicinity of Bering Strait was there any general reduction of tidal speed with depth (as contrasted to mean speed).

TRANSPORT

In addition to the eleven anchored current station sections across Bering Strait (Fig. 48), there are ten detailed sections entirely crossing the system from Alaska to Siberia that can be used in transport calculations, because runoff into the intermediate region is at least an order of magnitude less than the oceanic transport and thus does not contribute significantly to the total mass budget. Table 12 gives the values for transport through these sections determined by planimetry of the measured normal flow components. The measurements for the 14 July 1967 section (B, Fig. 2a) were not complete, due to malfunction of the current meter during occupation of the easternmost stations, so that the transport for

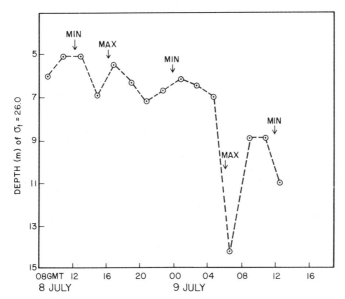

59. Depth of the isopycnal surface $\sigma_t = 26.0$ in Anadyr Strait, 8–9 July 1968 (Times of maximum and minimum currents below 10 m (cf. Figs. 57a, b) are indicated by the arrows.)

this section has been extrapolated by comparison with other sections and assuming the horizontal shear to be invariant.

Also included in Table 12 are the sectional mean normal wind stresses. The wind stress was calculated from $\tau_o = \rho C_D W^2$ where τ_o = stress (dynes cm^{-2}), $\rho C_D = 1.44 \times 10^{-6}$ (gm cm^{-3}), and W (cm sec^{-1}) is the wind speed component normal to the section. The averages were weighted by station separation.

The average north transport through the 21 sections was 1.2 Sv, but this value includes six sections between 15 and 22 July 1967 during what we believe to be anomalous conditions. The average excluding these sections is 1.7 Sv which, as four different years are included, would appear to be a reasonable estimate of normal summer transport.

We now inquire into the possible causes of the variations in transport. The values of normal wind stress and transport are shown in Fig. 60. There is not an overall good correlation between wind stress and transport, and yet during certain periods there is excellent correlation. This is particularly true during the four sections made the week of 3 to 7 August 1967. During the first section (4 August) there was no southerly set to the current anywhere in the section, the transport was >2 Sv toward the

TABLE 12
Transport (Sv; + north) and Mean Wind Stress (dynes cm^{-2}; + north) Through All Closed Sections Between Alaska-Siberia

Date		T, Sv	τ_o, dyne cm^{-2}	Location
1964	5–6 Aug	+ 1.4	− 0.43	Bering Str.
1967	13–14 Jul	+ 2.0*	+ 0.71	Bering Str. (B Fig. 2a)
	15 Jul	+ 0.3	+ 0.97	Shishmaref (C Fig. 2a)
	16–17 Jul	0	− 0.49	Pt. Hope (D Fig. 2a)
	18 Jul	+ 0.7	− 0.11	King Is. (A Fig. 2b)
	18–19 Jul	− 0.2	− 2.06	Bering Str. (B Fig. 2b)
	20–21 Jul	0	− 0.03	Shishmaref (C Fig. 2b)
	22 Jul	− 0.1	− 0.42	Pt. Hope (D Fig. 2b)
	3–4 Aug	+ 2.2	+ 0.13	Bering Str.
	4–5 Aug.	+ 1.6	− 0.69	Bering Str.
	5–6 Aug	+ 2.0	− 0.12	Bering Str.
	6–7 Aug	+ 1.4	− 1.54	Bering Str.
1968	9–11 Jul	+ 1.9	+ 0.09	Str. of Anadyr; St. Lawrence–C. Romanof (A Fig. 3)
	13–14 Jul	+ 1.2	− 0.23	C. Rodney (B Fig. 3)
	15 Jul	+ 1.4	0	King Is. (C Fig. 3)
	17–18 Jul	+ 1.5	+ 1.49	Bering Str. (D Fig. 3)
	18–19 Jul	+ 2.2	+ 0.94	Shishmaref (E Fig. 3)
1972	24–25 Jul	+ 1.7	− 0.12	Bering Str.
	27–28 Jul	+ 1.3	+ 0.07	C. Lisburne–C. Schmidt
1973	29–30 Sept	+ 1.9	+ 1.02	Bering Str.
	3–4 Oct	+ 1.4	+ 0.54	Bering Str.

*Extrapolated (see text).

north, and there was a small northerly wind stress. During the remaining sections, there was a northerly wind of varying strength with proportional amounts of southerly flow in the surface layer. The linear regression is

$$T = 2.06 + 0.48\, \tau_o$$

($r = 0.96$ is the correlation coefficient). Another grouping, of all remaining values >1 Sv (except 17 to 18 July 1968), is approximated by

$$T = 1.52 + 0.49\, \tau_o \quad (r = 0.72).$$

A third grouping, the six sections in the period 15 to 22 July 1967 in which all values were <1 Sv, is poorly correlated, with

$$T = 0.8 + 0.19\, \tau_o \quad (r = 0.56),$$

but if the value $T = +0.7$ (18 July) is omitted, then the correlation is excellent, with $r > 0.95$.

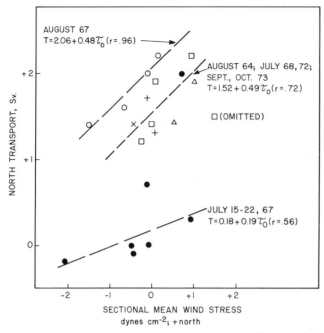

60. Correlation of Bering Strait transport with the sectional mean wind stress

An implication of these results is that local wind conditions significantly modify the mean flow. The wind stress locally appears to speed up or slow the water above the pycnocline. During conditions of normal northerly transport, the quantitative effect is about one-half Sverdrup of transport for each dyne cm^{-2} of wind stress applied. During the one period of anomalously low flows covered by our measurements, the wind effect appeared to be quantitatively less, being only about one-fifth Sv of transport per one dyne cm^{-2} of stress. The exact mechanism by which these effects are accomplished remains uncertain. Most likely the wind acts in conjunction with the land boundaries to force redistribution of the oceanographic pressure field. Certainly the current reversals are on the whole larger (by as much as one-half order of magnitude) than one would expect from a directly wind-driven pure drift current. Similarly, the observed transport changes are at least one-half order of magnitude larger than are appropriate to a fully-developed Ekman layer subjected to the corresponding wind stress (besides not being rotated relative to the N/S axis of the prevailing winds).*

It has been shown (Coachman and Aagaard 1966) that the major driving

*For example, for a change of wind stress of 1 dyne cm^{-2}, the Ekman transport across a normal 100 km cross-section changes by only 0.07 Sv at 65°N.

force for northward flow through Bering Strait is the sea surface sloping down to the north; for the section of 5 to 6 August 1964, the slope was calculated to be 2×10^{-6}. A north wind was blowing at the time, giving an average stress of -0.4 dyne cm^{-2}, so that we can estimate the transport of $+1.4$ Sv having been reduced about 0.2 Sv by the local wind effect. Therefore the slope of 2×10^{-6} is that which is associated with an average summer northerly transport of ~1.6 Sv. The normal condition is thus one in which sea level in the southern Chukchi Sea (in summer) is maintained about 0.5 m lower than in the northern Bering Sea. Thus, more or less continuously acting mechanisms must prevail regionally that tend to add water to the northern Bering Sea and/or remove water from the Chukchi Sea.

One possibility might be that as the majority of runoff takes place south of Bering Strait, this input would tend to maintain a higher sea level. Summer flow of the Yukon River at Ruby is of order 10^4 m^3 sec^{-1} (Fig. 26), which in turn is approximately 10 to 15% of the total runoff; if there was no flow out of the region between St. Lawrence and Bering Strait, approximately 5 days would be required for runoff to raise sea level 0.5 m. On the other hand, when there is little north flow through Bering Strait (as during 15 to 22 July 1967) sea level actually slopes down to the south (see Chap. 5). Under normal conditions, flow out of the region is not restricted and is at least an order of magnitude greater than the runoff, so that the fresh water input from land cannot be the basic cause of a higher sea level in the northern Bering Sea.

Runoff may, however, accumulate under certain specific conditions and subsequently augment the surface slope and increase the transport. For example, the correlations shown in Fig. 60 suggest that the northward transport during the first week in August 1967 would, under normal wind stress, have been 2 Sv or more, i.e., significantly higher than the normal summer transport of 1.6 Sv. Two weeks earlier occurred the anomalous period (15 to 22 July) when there was essentially no northward transport. The river flow all through June and early July 1967 had been normal, and at Ruby was $>10^4$ m^3 sec^{-1}. We therefore hypothesize that during the anomalous period the runoff was accumulated to the south, leading to above-normal amounts of water over the Bering Shelf region south of St. Lawrence Island. Then, on reestablishment of normal conditions under which water feeds into the system from the south, the excess water gave rise to a greater than normal pressure head which persisted for some time.

A major cause of variations in the sea level difference must lie in fluctuations of the regional wind distribution. It is also possible that the atmospheric pressure field may itself directly modify the oceanic pressure field. As an index to the regional atmospheric regime, we assembled (when available) daily synoptic (1200 Z) surface atmospheric pressures from the following stations throughout the region:

North of Bering Strait: Cape Schmidt, Cape Lisburne, Kotzebue
In Bering Strait: Cape Uelen
South of Bering Strait: Nome, Providence Bay, NE Cape (St. Lawrence)

The daily synoptic surface pressures at Nome (in the southeast) and Cape Schmidt (in the northwest) over the periods encompassing the transport measurements are plotted in Figure 61. Also shown are the transports and the difference in pressure between Nome and Cape Schmidt ($P_N - P_S$).

Certain qualitative correlations are evident:

(1) The incidences of very low transport (six cases between 15 and 22 July 1967) followed a precipitous fall of the surface pressure to the lowest values in the record. The pressure at Nome fell to <1000 mb and re-

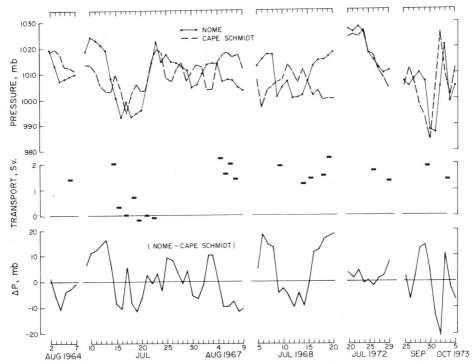

61. Daily (1200 Z) surface atmospheric pressures at Nome and Cape Schmidt and the difference (Nome minus Cape Schmidt) during July–August 1967 period when eleven cross-sections were occupied, and the same information during periods in 1964, 1968, 1972, and 1973 when eight cross-sections were occupied

mained <1000 mb for five days. At the same time, the pressure was higher in the northwest than in the southeast [$(P_N-P_S) \sim -10$ mb]. When these conditions temporarily reversed in the middle of the period [P_N went up to 1000 mb and (P_N-P_S) became positive on 17 July], one day later (18 to 19 July) the transport also went up temporarily (+0.7 Sv).

(2) The highest transports (>2 Sv on 19 July 1968, 14 July 1967, 4 August 1967) occurred, with a phase lag of 1 to 2 days, when the pressure at Nome was high (>1010 mb) and when (P_N-P_S) was strongly positive.

(3) Transports a little below average (13 to 14 July 1968, 15 July 1968, 5 to 6 August 1964) occurred during average pressure conditions, but followed, by 1 to 2 days, periods when the pressure difference (P_N-P_S) had been negative.

There thus appears to be a variety of interacting atmospheric conditions that cause variations in sea-surface slope through the region. In an effort to statistically quantize the transport pressure relationships a variety of correlations were investigated, among which were all possible single and multiple linear regressions on:

(a) east-west pressure gradients across the Chukchi Sea ($P_{Kotzebue}-P_{Schmidt}$) and the northern Bering Sea ($P_{Nome}-P_{Providence\ Bay}$);

(b) pressure at Nome and pressure at Cape Schmidt;

(c) pressure at Nome and pressure difference ($P_{Nome}-P_{Schmidt}$).

All correlations were tried with phase lags varying between one and three days.

The best correlation ($r = 0.86$) was for

$$T = 1.52 + 0.01\ (P_N-P_S) + 0.09\ (P_N-\bar{P})$$

where $\bar{P} = 1012$ mb is the average pressure over the record and all pressures are phase-shifted by one day previous to the section occupation.

Because the north-south pressure difference enters only weakly into the relationship (coefficient = 0.01), a single linear regression of transport on Nome pressure is nearly equally representative ($r = 0.79$):

$$T = 1.58 + 0.08\ (P_N - \bar{P}).$$

This correlation is shown in Fig. 62.

If only two of the 21 data points are deleted, the latter correlation is increased to $r = 0.88$. These are (1) 15 July 1967, when the transport was low (+0.3 Sv) while the pressure was still falling precipitously and had not yet reached values <1000 mb, and (2) 22 July 1967, when the transport remained low (−0.1 Sv) even though the pressure began rising two days before. In the first instance, no phase lag is indicated, i.e., using the

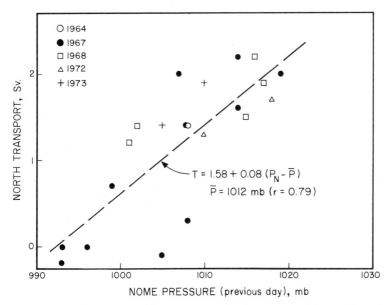

62. Correlation of transport with Nome atmospheric pressure of the previous day (1200 Z)

value of pressure for 15 July gives a better correlation, and in (2) a phase lag of about three days improves the correlation. Adjustments to these values are not warranted, however, as the system is certainly more complicated physically than what is represented by the simple correlation of transport with Nome pressure. Consider, for example, the changes between the four August 1967 sections; neither the Nome pressure nor the difference ($P_N - P_S$) reflect the variable transport sequence over the four days, whereas these changes are hindcasted quite well by the local wind stress variations (cf. Fig. 60). The available pressure data, lacking in sufficient synoptic detail, are totally inadequate to forecast these more minor modifications of the transport.

We conclude that transport through the region can be forecast to within about one-half Sv by the simple correlation with surface atmospheric pressure at Nome on the previous day. The reason for such a correlation must be that the Nome pressure reflects to considerable degree the regional wind regime that actually drives the flow, along with pressure variations that might also perturb the mean state. If the local wind stress is known, the calculation can apparently be further refined.

We interpret this regional regime as follows. The winds associated with average values of pressure at Nome have a northward component over the northern Bering Sea, which, acting on the meridionally bounded sea, sets up the oceanic pressure field which continually moves water northward. This transport is enhanced when the pressure to the northwest (Cape Schmidt) is low, giving a northward wind component over the Chukchi Sea also; and it is inhibited when the Cape Schmidt pressure is high. We note that low pressure at Cape Schmidt is in general associated with high pressure at Nome, and vice versa. When the pressure at Nome is much reduced, it is normally caused by an atmospheric low to the south moving eastward along the normal storm track, tending to move water out of the northern Bering Sea. This effect, if intense and prolonged as in the period 15 to 22 July 1967, locally lowers sea level to the south of Bering Strait below that to the north, and the rare occurrence of southerly transport results.

It also appears likely that major modifications of the mean northward motion can be driven directly by variations in the meridional atmospheric pressure gradient. During spring and summer 1973 we obtained two 120-day current records from 96 and 126 m depths in Barrow Canyon (Mountain et al., in press). Those currents were well correlated with the atmospheric pressure difference Barrow-Nome. The cross correlation was -0.74 for the pressure difference leading by one day, the negative correlation meaning that when the southward pressure gradient increased, the northward current speed decreased. It proved possible to model the flow variations quite well as perturbations induced by changes in the meridional atmospheric pressure gradient. By Fourier-decomposing the record of atmospheric pressure difference and using the amplitude and phase of each component in the solution to the perturbation momentum equation (with the eddy viscosity as a free parameter), a current record was generated that was extremely close to the observed currents (cross correlation of 0.83). In an attempt to relate these results to the present discussion of flow through Bering Strait and pressure at Nome, we undertook a review of pressure records from Barrow and Nome. While high pressure at Nome tends to correspond to low values of the pressure difference Barrow-Nome, the correlation isn't always good, generally ranging from -0.2 to -0.7. It is probable, however, that the available pressure observations are not really adequate for properly defining the meridional pressure gradient for Bering Strait. It is our belief that variations in the intensity of both the Bering Strait and Barrow Canyon flows tend to follow the meridional atmospheric pressure gradient over the area, with relatively high pressure to the north being associated with reduced northerly flow. The implication is that major variations in the flow pattern due to this mechanism are not generated in one area and propagate to another, but rather that the whole current system from Bering Strait to Barrow responds nearly simultaneously.

The local winds are also reflected to some degree in the Nome pressures: southerly winds with high pressure and northerly winds with low pressure. Thus, even though the system is obviously complicated, the various effects discussed interact in such a way that the Nome pressure gives a reasonable index to the total mechanism responsible for the flow.

The only measurements that give a suggestion of the transport through the region at times of the year other than summer are (1) those made from the *Northwind* in the vicinity of the Strait of Anadyr during the period 2 to 19 February 1968 (Fig. 44) and (2) the measurements made from the *Staten Island* in April 1969. Currents measured on this latter cruise from King Island northward through the eastern channel of Bering Strait between 21 to 24 April 1969 are depicted in Fig. 63.

To interpret the possible atmospheric effect on the *Northwind* measurements of February 1968, the pressure at Nome, the pressure difference ($P_N - P_S$), and the wind stress calculated from the wind observed during the hydrographic stations are shown in Fig. 64. For the latter calculation, positive stress was taken toward NE and negative toward SW;

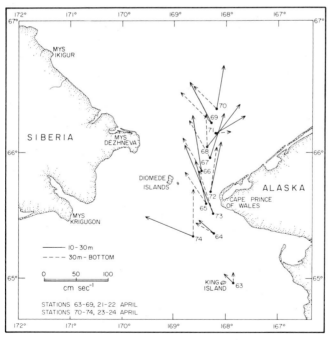

63. Averaged 10 to 30 m and 30 m: bottom layer currents measured in Bering Strait 21–24 April 1969 *Staten Island*

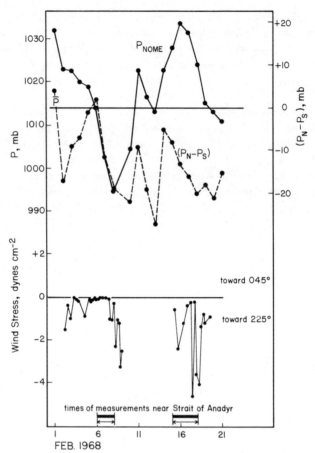

64. (Upper) Surface atmospheric pressure at Nome and pressure difference (Nome minus Cape Schmidt) during February 1968. (Lower) Northeast-southwest wind stress calculated from wind measurements aboard *Northwind* during February 1968 (Bars indicate periods of direct current measurements near Strait of Anadyr.)

the stations were occupied in the vicinity of the Strait of Anadyr only during 6 to 8 February and 15 to 17 February. During the first period, the Nome pressure was falling steeply to values <1000 mb on 7 February, and a large negative pressure difference developed concurrently. In addition, a strong wind stress to the SW began on 7 February. These were thus conditions suitable for a reduced northerly flow, or even a reversal. During the second period the pressure at Nome had risen to above average values by 14 February, but the pressure was still much higher at Cape Schmidt. In addition, this was a period of very strong southwesterly-directed wind stress (observations were not made on 13 or 14 February, but on 17 February the SW stress was >4 dynes cm^{-2}). Though the data are not conclusive, we believe the conditions prevailing at the time of the February 1968 current observations to have been conducive to reduced northerly (or reversed) flow, supporting our previous conclusions of the February measurements as being anomalous.

The current measurements in Bering Strait during April 1969 (Fig. 63) were relatively swift (50 to 100 cm sec^{-1}). The velocities at stations 65 and 72 in the eastern channel are quite similar in direction and magnitude to those from nearly the same locations during 17 to 18 July 1968 (Fig. 47) and 24 to 25 July 1972 (Fig. 84). The north transports through the section at these times were 1.5 Sv and 1.7 Sv, respectively. As the horizontal shear in the Bering Strait section seems to be relatively invariant, we conclude that during 21 to 24 April 1969 there is good evidence that the transport was north >1.5 Sv. Now, using $T = 1.58 + .08 (P_N - \bar{P})$, transports would be predicted as follows:

April	T, Sv
21	+1.5
22	+2.5
23	+2.5
24	+2.2

This trend is compatible with the current scheme depicted in Fig. 63.

Soviet oceanographers have for many years reported that transport in winter is only one-third to one-fourth that in summer, an idea which seems to have originated with Maksimov (1945), but to our knowledge no supporting data are available. Both Fedorova and Yankina (1964) and Antonov (1968) have shown a relatively smooth annual cycle with a minimum in March (0.4 Sv) and maximum in August (1.6 Sv).

Our measured transports are in close agreement with the Soviet work for summer. At other times of the year, however, the evidence we have assembled suggest flows not materially different, on the average, than are encountered during summer. Current measurements in March 1967 north of Bering Strait showed the mean flow to be the same as in summer (Coachman and Tripp 1970), and likewise with the April 1969 measurements cited above. The February 1968 measurements suggested anomalous reduced or reversed flow in Anadyr Strait for a time, but atmospheric conditions during the remainder of February seemed appropriate to normal north transport. For example, assuming the above transport-pressure relationship, the average February 1968 transport calculated from the daily pressure values would have been +1.5 Sv. For comparison, average transports for summer 1968 (29 June to 21 July) and summer 1967 (9 July to 9 August) were +1.6 Sv and +1.3 Sv, respectively.

We conclude that there is no satisfactory evidence for an annual cycle of transport through Bering Strait with low mean values (<1 Sv) in winter and high values in summer. Rather, transport can vary by a factor of two on much shorter time scales (~1 week), and these variations tend to cancel when averaged over a month. Thus, the mean monthly transport throughout the year likely ranges between 1 and 2 Sv, which is also the value of mean annual transport.

4

CHUKCHI SEA

INTRODUCTION

The Bering Sea water masses advected north through Bering Strait traverse the Chukchi Sea, a distance of some 700 km, before entering the Arctic Ocean. The water undergoes considerable modification en route, but the water masses still can be recognized as a significant contribution to the subsurface (pycnocline) layer of the Arctic Ocean (Coachman and Barnes 1961).

The Chukchi is a continental shelf sea, but its floor has somewhat greater relief than the northern Bering (Fig. 1). The north central part is dominated by Herald Shoal, with depths <30 m. South of Herald Shoal and west of Pt. Hope, the depth is nearly 60 m; this is the inner end of a long submarine valley, an extension of Herald Canyon which indents the shelf north of Herald Island, and it seems to terminate in Kotzebue Sound. Near Herald Island, canyon depths are ~90 m.

On the east side of Herald Shoal there is a depression about 50 m deep. It does not appear to be continuous with any submarine canyon feature, since depths <50 m are encountered both north and south.

In the northeast Chukchi, Barrow Canyon, with depths >50 m as far south as Cape Franklin, indents the shelf parallel and close to the Alaskan coast. It shows a sectional asymmetry, with the steepest side next to the coast.

The Chukchi Sea connects with the East-Siberian Sea to the west through Long Strait, about 135 km wide with a sill depth of ~45 m.

Two large spitlike shoals extend northward from the westward-projecting peninsulas, Seward and Pt. Hope/Cape Lisburne. These spits are composed of sediments transported from south of the promontories by the prevailing currents (Creager and McManus 1966), and they influence the flow field significantly.

The most comprehensive analysis of the Chukchi Sea regime to date has been that of Aagaard (1964). It was based largely on the 1962 October *Northwind* cruise, the only quasi-synoptic survey of the entire sea ever accomplished; but it also utilized other data, including the 1922 *Maud* stations from the northwestern Chukchi, the *Brown Bear* work of 1960 over the eastern portion of the sea as far north as Icy Cape, and the 1963 *Northwind* sections along the Siberian coast to Long Strait.

The analysis was based on three regional water mass groupings, the Alaskan Coastal, Siberian Coastal, and central and subsurface waters. The Alaskan Coastal Water, with salinities $<31°/_{00}$, flows north along the eastern side of the sea. In summer this water is warm, with temperatures characteristically between 10 and 15°C, and with large gradients of temperature (decreasing westward) and salinity (increasing westward) (cf. Fig. 24). This situation prevailed in October 1962, but the temperatures were much lower. The salinity values were slightly lower in early October than on previous summer cruises, and slightly higher in late October. The temperature and salinity gradients were also less in early October, and had markedly diminished by the end of the month. The loss of heat in the southern Chukchi Sea during autumn was estimated to be 30% due to local surface flux and 70% due to advection of colder water from the south.

The flow of the Alaskan Coastal Water northward along the eastern side of the sea (cf. Fig. 46) is to a marked degree parallel to the bottom contours, in accord with the conclusion of Fleming and Heggarty (1966). Thus, the flow is northward from Bering Strait and then, north of the Cape Prince of Wales spit (approx. 60 to 80 km north of the strait), it curves northeast and east toward Kotzebue Sound. The degree of penetration into Kotzebue Sound seems to vary; in summer 1959 it was farther than either summer 1960 or October 1962. The flow converges toward Pt. Hope where the water is accelerated. Near Pt. Hope the flow bifurcates, one part setting northwest along the south side of Herald Shoal and the other northeast along the Alaskan shore.

Below the low-salinity Alaskan Coastal Water in autumn 1962 lay more saline water (31.5 to $32.4°/_{00}$), which on the whole was warmer (2°C) than the surface water. This water also to a marked degree tended to follow the bottom contours, i.e., proceeding toward Pt. Hope, but its core was displaced to the west of the low-salinity surface water. Northwest of Pt. Hope the flow of subsurface water likewise appeared to bifurcate, with one part setting northwest south of Herald Shoal and the other northeast. (When winds were from the northeast quadrant, the surface flow northeast along the Alaskan coast was slowed or reversed, while it appeared that the deeper flow was less affected.)

Siberian Coastal Water during the 1962 *Northwind* cruise was colder than −1°C and had salinities $<30.5°/_{00}$, and it flowed southeast along the Siberian coast. The distance this coastal current penetrates toward Bering

Strait varies; in October 1962 it reached within 80 km of Cape Serdze-Kamen, but under sustained west or northwest winds it may reach Cape Dezhneva. The lateral extent of the Siberian water from the coast is relatively great in Long Strait and then narrows toward the point of its eastern penetration.

Temperatures and salinities in the Siberian Coastal Water vary widely both annually and seasonally, but the main characteristic is a broad range of relatively low salinity and temperature. Various possible sources of the salinity deficit were examined by Aagaard, who concluded that they must be in the East-Siberian Sea, the fresh water being supplied mainly by the Kolyma and Indigirka rivers. Local ice melt also reduces the salinity, particularly in early summer.

The central Chukchi Sea water beneath the surface layer showed characteristic temperatures of 1°C and salinities of 32.9°/$_{00}$, but the most striking feature was a large pool of oxygen-deficient bottom water in the south-central region (Fig. 65). This water must have been resident over summer and autumn, with no advection through its location. Assuming that the resident water was advected in during June with an O_2 saturation of 80% and that the effects of photosynthesis and diffusion could be neglected, an O_2 utilization rate of 11 ml/l/yr was estimated (see discussion below in *Water Masses*).

The surface waters of the central and northern parts of the sea ranged in salinity between 31 and 32.5°/$_{00}$, with a tendency to increase toward the

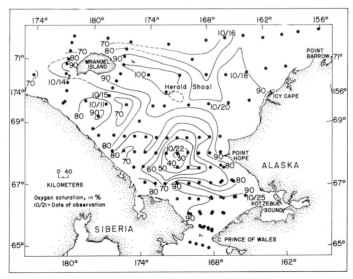

65. Percent O_2 saturation of bottom water, 1962 *Northwind* (from Aagaard 1964)

north. In late October, surface temperatures were lower than at subsurface levels due to fall cooling, but in summer the warmest values are observed at the surface. There was evidence of vertical mixing adjacent to Herald Shoal.

Along the Siberian and Alaskan coasts, there appeared to be a near-bottom encroachment of cold, high-salinity ($>33°/_{00}$) water. In Long Strait, this seemed to be associated with the current carrying Siberian Coastal Water. Even more saline water (33.9 to $34.3°/_{00}$), with temperatures of -0.7 to $-0.9°C$, was observed at depth southwest of Pt. Barrow. It was concluded that this was Arctic Ocean water from >100 m depth, and that its presence was associated with Barrow Canyon.

Data obtained since 1964 (see Table 1) include the sections of closely-spaced anchored stations made from the *Northwind* in July 1967 (Figs. 2a, b); a detailed hydrographic survey during September to October 1970 of a triangular area lying north and west of Cape Lisburne and Icy Cape, including some current measurements (*Websec-70*); and the late July 1972 cruise of the *Oshoro Maru* (Fig. 66). A four-day current record was obtained in central Kotzebue Sound in March 1968 (Coachman and Tripp 1970), and during August 1966 about 440 current measurements were made in Long Strait (Coachman and Rankin 1968).

WATER MASSES

There are three water masses in Bering Strait (arranged laterally from east to west: Alaskan Coastal, Bering Shelf, Anadyr), differentiable on the basis of salinity. Within a short distance north of the strait the water masses reduce to two through combination of the Anadyr and Bering Shelf waters. Figure 67 shows the 15 July Shishmaref section of 1967 *Northwind* (C Fig. 2a) and Fig. 68 shows the Pt. Hope section (D Fig. 2a). The Anadyr and Bering Shelf water masses, originally separated by 0.2 to $0.3°/_{00}$ in the salinities of their deeper layers, have blended laterally such that the distinguishing salinity separation is lost. However, the distinct separation between these water masses and the lower-salinity Alaskan Coastal Water is not altered.

Thus, in effect, a new water mass is created just north of Bering Strait. It has a salinity range encompassing those of the parent Bering Shelf and Anadyr water masses, which in summer 1967 was 32.6 to $33.2°/_{00}$ (a year of saline waters, cf. Fig. 35), whereas in 1963 the range was 32.5 to $33.05°/_{00}$ (*Northwind* data) and in 1972, 32.3 to $33°/_{00}$ (years of less saline water). This new water mass we shall call Bering Sea Water, and it dominates the central and western part of the southern Chukchi Sea. It exhibits about

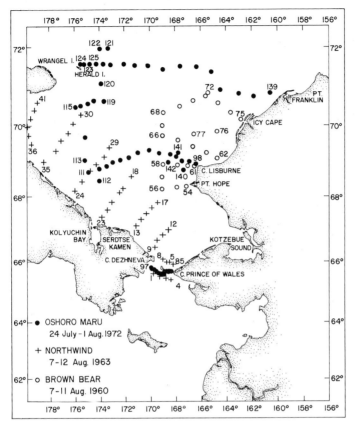

66. Station locations in Chukchi Sea of *Brown Bear* (7–11 August 1960), *Northwind* (7–12 August 1963), and *Oshoro Maru* (24 July–1 August 1972)

the same temperatures as the parent water masses in Bering Strait, which in summer have minimum values of about 2°C.

The sharp demarcation between Bering Sea Water and Alaskan Coastal Water is reinforced by a further low-salinity contribution to Alaskan Coastal Water of water from Kotzebue Sound. This water has the same broad range of low salinity and (in summer) high temperatures as Alaskan Coastal Water, and seems to join in the general northward flow toward Pt. Hope. The slightly lower near-bottom temperatures at stations 40 and 41 in the Pt. Hope section, compared to any observed in Bering Strait (Fig. 68), may be due to Kotzebue water.

We believe that some of the water from Kotzebue Sound moves southwest along the north shore of the Seward Peninsula, approximately

116 • BERING STRAIT: The Regional Physical Oceanography

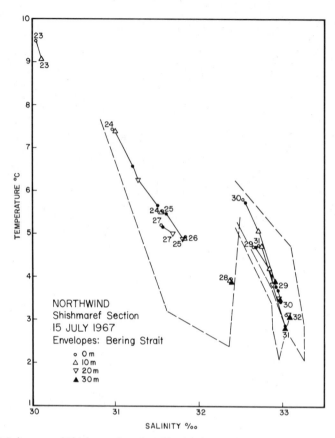

67. T-S diagram of Shishmaref section (Fig. 2 left) 15 July 1967 (*Northwind*) and T-S envelopes of water masses in Bering Strait (Fig. 41): station numbers from east to west, 23 to 32.

as far as Shishmaref, before joining the general northward flow. In the 18 to 19 July 1968 Shishmaref section (Fig. 69), the three easternmost stations (68 to 70) showed water definitely less saline than Alaskan Coastal Water as observed in Bering Strait. Current measurements near Shishmaref (cf. Fig. 46) show flow in the area to have a west component. Thus, the data suggest an anticyclonic eddy in the lee of the Cape Prince of Wales spit and lying south of the main northward flow.

The penetration of Siberian Coastal Water as far as Bering Strait can be questioned. In none of the detailed sections, in which stations were made to within 25 km of the Siberian coast north of the strait, is there any sign of water which can be interpreted as having derived from Siberian Coastal Water. Even during the period of flow reversal during July 1967,

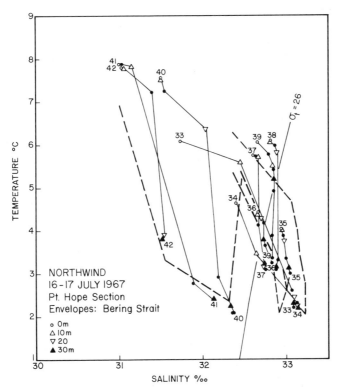

68. T-S diagram of Pt. Hope section (Fig. 2 left) 16–17 July 1967 (*Northwind*) and T-S envelopes of water masses in Bering Strait (Fig. 41): station numbers from east to west, 33 to 42

the Pt. Hope section (Fig. 70) showed no cold low-salinity water. Station 83, closest to Siberia, did have bottom water of slightly higher salinity and nearly one degree colder than Bering Sea Water. However, the characteristics suggest this water to be resident Chukchi Sea bottom water (see below), which has moved southeast of its normal position during the unusual southerly flow event.

The possibility remains that cold, low-salinity Siberian Coastal Water on occasion penetrates to Bering Strait <25 km from shore. The only Soviet data available are due to Ratmanov in 1932, 1933, and he observed net south flow at the station closest to Cape Dezhneva in 1933. A close examination of his published temperature and salinity distributions at 10 m, 25 m, and near-bottom for summers of both 1932 and 1933 shows no water near Bering Strait with characteristics that cannot be accounted for by a continuity of water masses with the northern Bering Sea. It thus appears that sufficient water is supplied to the southern Chukchi Sea

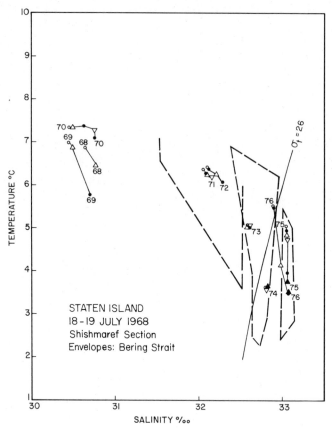

69. T-S diagram of Shishmaref section (Fig. 3) 18–19 July 1968 (*Staten Island*) and T-S envelopes of water masses in Bering Strait 17–18 July 1968 (Fig. 6): station numbers from east to west, 68 to 76

from the south to prevent Siberian Coastal Water from normally penetrating to Bering Strait. As suggested earlier, the rare and relatively short flow reversals only move the Bering Sea Water south temporarily; it would take a prolonged flow reversal to completely flush the system, and such an occurrence has yet to be documented.

The transit and modification of water masses through the central and northern Chukchi is best followed during midsummer when temperature and salinity ranges are greatest and thus water mass differentiation and interactions are most clearly delineated. Data from three midsummer cruises, 1972 *Oshoro Maru*, 1960 *Brown Bear*, and 1963 *Northwind* (Fig. 66) were compared, and temperature and salinity values were found to be compatible. Average temperature and salinity distributions within the

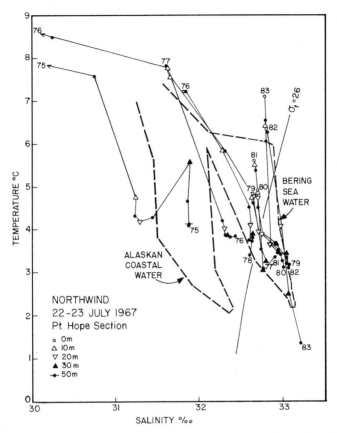

70. T-S diagram of Pt. Hope section 22–23 July 1967 (*Northwind*; Fig. 2 right) and T-S envelopes for water masses in the Pt. Hope section on 16–17 July 1967 (Fig. 68): station numbers from east to west, 75 to 83

surface layer (0 to 10 m) based on these data are shown in Figs. 71 and 72, respectively.

The warmest water lies on the east side of the system south of Pt. Hope, and shows the intimate connection with Kotzebue Sound. Northwest of Pt. Hope there is a marked branching. The western limb passes south and west of Herald Shoal and then extends northward into the Arctic Ocean, coincident with Herald submarine canyon (cf. Fig. 1). The eastern limb of warm water extends northeast from Cape Lisburne approximately parallel to the Alaskan shore, and is nearly severed off Icy Cape by a tongue of cold water from the northwest.

The salinity distribution shows that the low-salinity Alaskan Coastal Water on the eastern side is fed by an even somewhat fresher contribu-

120 • **BERING STRAIT: The Regional Physical Oceanography**

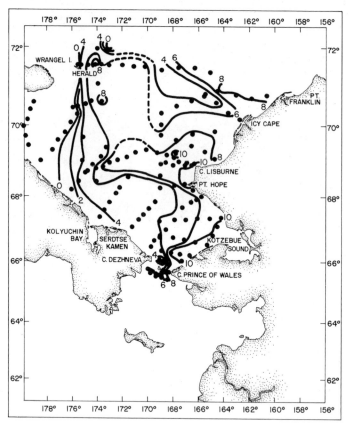

71. Average temperature °C of 0 to 10-m layer during midsummer, based on stations of Fig. 66

tion from Kotzebue Sound, which mixture dominates the northeastward-moving branch north of Cape Lisburne. The northwest branch contains the higher salinity (>31°/$_{00}$) water from the south.

Near Wrangel Island, Long Strait, and the Siberian coast lies water of markedly lower salinity which is also very cold, the Siberian Coastal Water described by Aagaard. There is a marked frontal zone between the two water masses, the one of Bering Sea origin and the other from the East Siberian Sea. Thus, the Bering Sea Water progressing northward through the central Chukchi Sea constitutes a tongue of relatively high surface salinity, with less saline water both on the west and east sides. The western boundary of the tongue is marked by strong gradients of temperature and salinity, while the transition on the east side toward low salinity and high temperature is less marked.

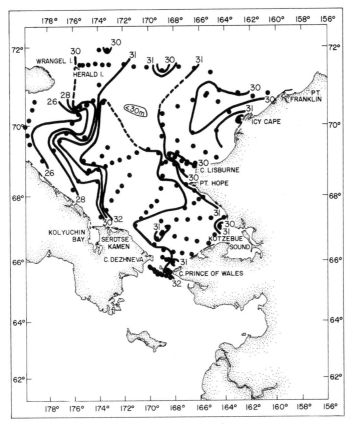

72. Average salinity ‰ of 0 to 10-m layer during midsummer, based on stations of Fig. 66.

The Bering Sea Water formed in the southern Chukchi Sea from Anadyr and Bering Shelf waters had a salinity range from about 32.2 to 33‰ during these cruises. Water in this range lies beneath the surface layer, and, to provide an initial view of property distributions in the deeper layer in midsummer, mean temperature and salinity for the layer 30 m to the bottom are shown in Figs. 73 and 74, respectively. The patterns of temperature and salinity are, in general, similar to those of the surface layer. The less saline water stays along the Alaskan shore, within the northeastern branch of the bifurcated flow. It carries with it the highest temperatures. The more saline fraction of the relatively warm water predominates over the central Chukchi Sea, and its distribution indicates a movement into the Arctic Ocean west of Herald Shoal, as was also the case in the upper layer. However, in the southern and central

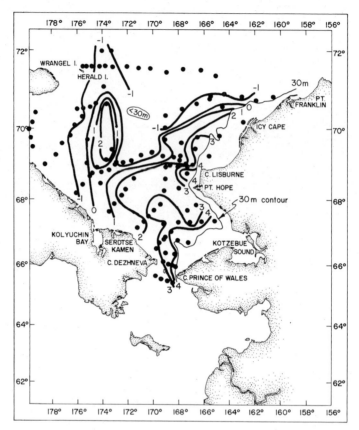

73. Average temperature of bottom (>30 m) layer in midsummer, stations of Fig. 66

region, the core of this water defined by the tongue-like isotherms appears to be shifted significantly to the west of the core of the upper layer.

Another similarity with the surface layer distributions is the evidence of a Siberian Coastal Water intrusion from the East Siberian Sea along the Siberian coast, as indicated by the cold water with salinities less than those of the Bering Sea Water (<32.2‰). There is also evidence for intrusions into the Chukchi of waters with salinities greater than Bering Sea Water (>33‰ during these cruises). A possible source is the East Siberian Sea via Long Strait. The intruding cold (<−1°C) and saline water appears to extend to the north around the east side of Wrangel Island, and thus borders on the west the Bering Sea Water contiguous with Herald submarine canyon. The second saline intrusion is associated with the inner end of Barrow submarine canyon, where the three stations

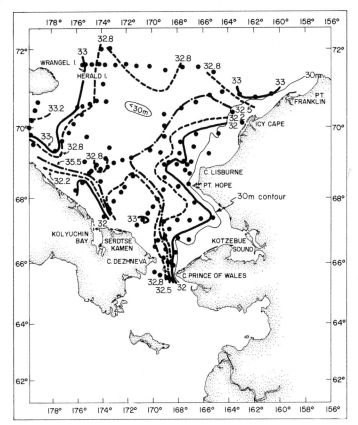

74. Average salinity of bottom (>30 m) layer in midsummer, stations of Fig. 66

north of Icy Cape have average salinities below 30 m of more than 33‰. These data thus confirm Aagaard's observation of Arctic Ocean water at depth in Barrow Canyon.

Further confirmation of the intrusion of Arctic Ocean water into Barrow Canyon was recently obtained from two current meter and temperature records (Mountain et al., in press). Instruments were moored 20 km NW of Pt. Barrow in 150 m of water on the canyon axis at depths of 96 and 126 m for the period 17 April to 17 August 1973.

The records show that for the majority of the time water near the freezing point flowed out-canyon toward the deep Canadian Basin. Late on 20 May the flow reversed and reached onshore speeds of 40 cm sec^{-1}. Twenty hours after the current reversal, the temperature at the lower current meter increased rapidly, reached +0.2°C, and remained at that high level for over 2 days. During the latter part of this period of elevated temperatures, the direction of flow at the lower meter again reversed to out-canyon; but about one-half day of offshore flow was required before the temperature again dropped to $<-1°C$.

The relatively warm water recorded at the lower meter was clearly Atlantic Water of the Arctic Ocean, for which the 0.2°C isotherm in the Canadian Basin occurs between 300 and 350 m. Such prolonged periods of onshore flow are common in the records, although temperatures as high as +0.2°C were reached only once during the four months of observation.

Oxygen determinations were made on two of the three summer cruises, and O_2 percent saturation of the near-bottom water is shown in Fig. 75. The distribution conforms qualitatively to that for fall 1962 (Fig. 65). The lower layers of the Alaskan Coastal Water are close to saturation,

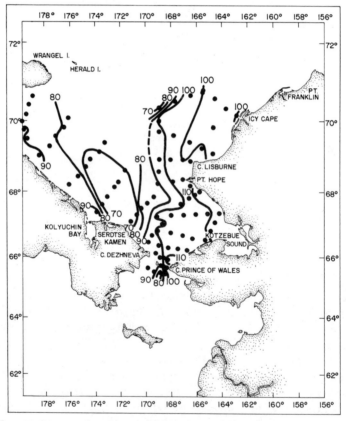

75. Percent O_2 saturation in near-bottom layer, midsummer (1960 *Brown Bear*, 1963 *Northwind*, see Fig. 66)

while the main body of Bering Sea Water comprises water with 80 to 100% O_2 saturation. There is a region of minimum O_2 in the central Chukchi in midsummer, though when these data were taken it occupied a more westerly position than during fall 1962, and O_2 values were not as low.

We conclude that Aagaard's depiction of water mass distributions is reasonable, and that the general features of these distributions are the same during summer and fall. With the additional information now available, certain aspects of the distributions can be further elucidated.

The earlier conclusion that water masses in this area seem constrained to follow bottom contours (Fleming and Heggarty 1966) is strongly reinforced. In the southern Chukchi the eastward bending of isotherms and isohalines into Kotzebue Sound conforms closely with the 30- and 40-m contours of the most recent bathymetric charts (Fig. 1). North of Cape Lisburne, where the 30- and 40-m contours diverge from the coast, the core of the Alaskan Coastal Water similarly diverges, and an anticyclonic eddy is formed in the lee of the Pt. Hope-Cape Lisburne peninsula. The eastern side of the eddy is shown by the southward trend of the 8°C isotherm (Fig. 71); the current measurements made during *Websec-70* (Ingham et al. 1972) also support the presence of a semipermanent eddy in this location.

The coincidence with Herald Canyon of the northwestward branch of the bifurcated flow, and of the northeastward branch with Barrow Canyon, is further evidence for strong bathymetric influence. Likewise, the apparent movement of Arctic Ocean Water toward the south in the north-central region (Fig. 73) appears to coincide with the somewhat greater depths east of Herald Shoal.

The presence in the south-central Chukchi of a bottom water resident during summer is confirmed. There is not a direct connection between this water and deep water from the East Siberian Sea, for the latter is both more saline and much colder. The temperature and salinity characteristics of the resident water are instead very close to those of the Bering Sea Water, which appears to circumnavigate the resident water along the east and north sides, and to some degree also tends to overlie the resident water. The resident water is bordered along the west side by the Siberian Coastal Water.

The striking feature of the resident water is its low oxygen concentration, though values measured in summer (Fig. 75) were not as low as in autumn (Fig. 65). The thickness of the low-oxygen layer is only about 10 m. As pointed out by Aagaard (1964) there can be no advection through the minimum, so that the lowest O_2 values are an indicator of the approximate center of the gyre. Thus, the position of the gyre varies; in the summer data presented here, the residual core lay farther west than during autumn 1962. Quite conceivably, the exact position could shift quite frequently during the summer as a result of bulk advective displacements which, however, prove insufficient to flush the water out of the Chukchi.

We hypothesize that the resident Chukchi bottom water is formed in the same manner as bottom water in the East Siberian Sea (Sverdrup 1929) and discussed for the Bering Sea in Chap. 2. The formation of a resident water undoubtedly occurs over much of the Chukchi Sea each winter, as on the northern Bering Sea shelf. Subsequently, the bottom water remains in residence by inhabiting the center of a cyclonic circulation.

The minimum O_2 concentration was about 5 ml/l, and in autumn 3 ml/l. Saturated water of $-1.8°C$ and $33°/_{00}$ would have 8.4 ml/l. If isolation of the resident water begins in April, and the effects of diffusion and photosynthesis are negligible, a minimum estimate of the mean O_2 utilization rate for central Chukchi bottom water between April and midsummer is 9 ml/l-yr, and between midsummer and October 9.6 ml/l/yr. These rates are somewhat less than that estimated by Aagaard (11 ml/l/yr) because they are based on a different assumption regarding bottom water formation. However, they are high rates when compared with those calculated for such bodies of water as Puget Sound (6 ml/l/yr; Barnes and Collias 1958).

In contrast, the same calculations were made for the East Siberian Sea bottom water evidenced in summer and autumn intruding into the Chukchi through Long Strait. The minimum rate for spring/summer is 2.4 ml/l/yr, and for summer/autumn 5 ml/l/yr. We conclude that the oxygen demand within the resident bottom water of the central Chukchi Sea is unusually high; the subject demands further study.

Of the water entering the Chukchi Sea through Long Strait from the East Siberian Sea, the cold but less saline fraction penetrates a varying extent southeast along the Siberian coast as a surface flow. Neither it nor the very cold ($-1.8°C$) and saline ($>33°/_{00}$) fraction, East Siberian Sea bottom water, normally appears to move into the central Chukchi Sea. A strong frontal zone demarks the boundary between these waters and Bering Sea Water of the central Chukchi. In summer the front is seen in both the upper and lower layers as a large temperature gradient, and in the upper layer as a strong salinity gradient. The general position of the front is from the vicinity of Wrangel Island southeast toward Kolyuchin Bay, but its exact location varies with the direction of net flow through Long Strait and with the degree of penetration of East Siberian Sea waters (see below).

Penetration of Arctic Ocean water into the northern reaches of the

Chukchi Sea also appears to be confirmed. In part this is effected by a southward flow north of Herald Shoal, which subsequently turns east to join the northeastward flow along the Alaskan coast. In the summer data this water was slightly more saline (~32.8°/$_{00}$) and noticeably colder (<-1°C) than Bering Sea Water. In autumn 1962, the salinity values were lower, but the contrast with Bering Sea Water was the same, i.e., the Arctic Ocean contribution was more saline than the Bering Sea Water.

The presence at depth in Barrow Canyon of saline (>33°/$_{00}$) water, which, even though not as saline as during autumn 1962 (>33.5°/$_{00}$) still showed no contiguity with other waters in the Chukchi, can only be attributed to inflow from the Arctic Ocean, as pointed out by Aagaard. The tentative hypothesis was that submarine canyons behave as submerged estuaries, in which offshore water may be brought in along the bottoms of the canyons as the surface water moves seaward. The evidence at the present time is that the deep water intrusion associated with Barrow Canyon during autumn 1962 was not a unique occurrence, as witnessed by the 1973 measurements. Such intrusion might also be anticipated in Herald Canyon, but there is no evidence in these data of Arctic Ocean water north of Herald Island at 72°N. However, the position of these stations is at least 275 km south of the mouth of Herald Canyon in the Canadian Basin continental slope; while the observations west of Pt. Franklin are less than 200 km from the mouth of Barrow Canyon. Examination of 1922 *Maud* data (Sverdrup 1929) shows that at station 14, at 73°N (i.e., 110 km farther north than 1972 *Oshoro Maru* stations 121, 122), the water deeper than 80 m was >33°/$_{00}$ (depth to bottom = 100 m), whereas at station 9 (70 km farther south) the water column was <33°/$_{00}$. We conclude that there is reasonable evidence that the Herald and Barrow canyons provide avenues for ingress of Arctic Ocean water to the northern Chukchi Sea.

Bering Sea Water

The analytic separation of Bering Sea and Alaskan Coastal waters is based on the two salinity modes. In the summer data, Bering Sea Water was in the range 32.2 to 33.0°/$_{00}$, a σ_t range of ~25.9 to 26.4 (cf. Fig. 7), while Alaskan Coastal Water had a broader range of salinity <32.1°/$_{00}$. As these water masses dominate the Chukchi en route to the Arctic Ocean, we shall now follow their modification as they interact with waters from the East Siberian Sea, resident bottom water, and intrusions from the north.

Figure 76 shows T-S envelopes of 1963 *Northwind* sections made normal to the Siberian coast (for station locations see Fig. 66). The envelope for stations 13 to 17, approximately coincident with the envelopes of the

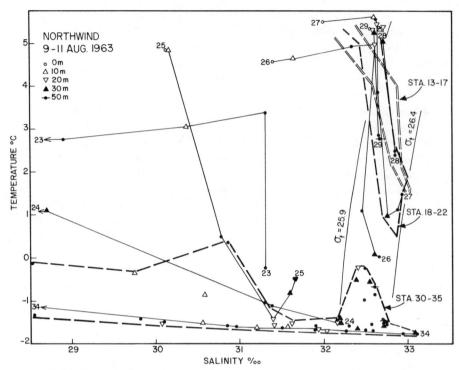

76. T-S diagrams of 1963 *Northwind* sections normal to Siberian coast; for station locations see Fig. 66

Pt. Hope sections from the 1967 *Northwind* cruise (Figs. 68 and 70), demonstrates again, at this location downstream from Bering Strait, the complete blending of Anadyr and Bering Shelf waters to form the Bering Sea Water. The stations did not extend far enough toward Pt. Hope to include Alaskan Coastal Water.

Northwestward from Pt. Hope, the modification observed in the *Northwind* sections was primarily a progressive vertical mixing of Bering Sea Water with water of the same density, but colder ($<-1.5°C$) and slightly less saline (32.2 to 32.8‰), such that the mixed water remained within the original isopycnal band. The water with which the Bering Sea Water mixes is not that of the Siberian coastal current, which was first encountered at station 23 and in the next section westward at stations 24, 25; the low salinity values show that no significant interaction with the Bering Sea Water was taking place. Likewise, deep water from the East Siberian Sea, with high salinities ($>33‰$) and near-freezing temperatures ($<-1.5°C$), does not appear until station 34, where there is a marked layering, with the now considerably cooled Bering Sea Water overlying the Siberian Sea water.

Instead the water with which the Bering Sea Water is mixing must be Chukchi Sea water resident from the previous winter. In the south-central Chukchi, however, vertical transfer of salt and heat has been sufficiently great that by midsummer no vestige of the winter conditions remains. Even though the T-S values of the deeper water have been altered through interaction with Bering Sea Water, the dissolved oxygen content of the near-bottom layer is extremely low, as previously noted. The implication is that the T-S mixing is accomplished quite early in the season and/or the oxygen utilization rates are extraordinarily high.

The core of Bering Sea Water is evidenced at stations 28 and 29 (Fig. 76) by the highest values of deep-water temperature, about 2.5°C in this location during 1963. These stations are closest to the central Chukchi Sea, and it appears that the core of Bering Sea Water had passed north of sections 13 to 17 and 18 to 23 in its northwestward progress. This suggests that during summer the last remnants of the resident water are not widespread over the whole central Chukchi Sea, but rather occupy a more restricted area in the south-central region. This is of course also the implication of the distribution of low-oxygen water. There is only a limited supply of resident water, which in summer is not being renewed.

The Bering Sea Water does not extend south to the Siberian coast, as there is a sharp frontal zone between the Bering Sea-resident water mixtures and the Siberian Coastal Water. The westernmost extent of Bering Sea Water in 1963 was section 30 to 35, where it fronted against East Siberian Sea deep water as well as some upper water of the Siberian Coastal type. The frontal zone on the west showed a markedly layered character, with East Siberian deep water at the bottom. The front did not extend west into Long Strait, as all stations in section 36 to 41 showed T-S curves similar to station 34 (Fig. 76). It therefore appears that at approximately 175 to 176°W the core of Bering Sea Water had turned north toward Herald Canyon.

The 1972 *Oshoro Maru* Cape Lisburne section was oriented essentially east-west across the central Chukchi Sea (stations 98 to 111; Fig. 66) and the T-S data are plotted in Fig. 77. Alaskan Coastal Water, in the salinity band 31 to $32.2^o/_{oo}$ and with a minimum T of ~1°C, is clearly distinguished in the stations near Cape Lisburne (98 to 102). Bering Sea Water was observed at the 3 westernmost stations (109 to 111), with the core at station 110 identifiable by the warmest deep water. It appears that the midsummer conditions were not identical in 1963 and 1972; in 1972 maximum Bering Sea Water core temperatures were lower than in 1963, and the core was located somewhat farther south (perhaps 30 to 40 km).

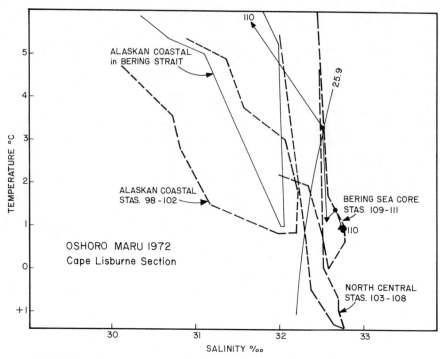

77. T-S envelopes of water masses in Cape Lisburne section 1972 *Oshoro Maru* (see Fig. 66)

At the location in the west-central Chukchi of stations 110 (1972) and 28 to 29 (1963) (~174°W), where the Bering Sea Water appears to turn northward toward Herald Canyon, the distance of travel from Bering Strait is about 450 to 500 km. There had been no significant modification in transit of the salinity of the Bering Sea Water, but it had been cooled. An estimate of the amount of cooling can be had by comparing deep core temperatures with those near Bering Strait. In 1963, of the 4 stations occupied north of the strait (5 to 8, Fig. 66), station 8 showed the highest salinity deep water, with a corresponding temperature of 3°C. This suggests a cooling in transit of about 0.5°C. In 1972, the core temperature was about 1.5°C in Bering Strait (Fig. 7), also suggesting a cooling of ~0.5°C. On the south side of the core the amount of cooling had been greater because of greater mixing with the cold resident water.

On the north side of the core of Bering Sea Water in the central Chukchi lay water in 1972 with the same T, S characteristics as the original southerly resident water, that is, cold and in approximately the same salinity band (32.2 to 32.8‰). This water was sampled by the Cape Lisburne section and is identified as the "north central" envelope of stations 103 to 108 (Fig. 77). It seems most likely this is also residual winter-formed Chukchi Sea water, but that it has been part of a more vigorous circulation than the south-central resident water. For example, it could have been brought from north of Herald Shoal into the central region by the southerly current which apparently exists east of the shoal. No significant mixing between the Bering Sea Water and the residual Chukchi water to the north appeared to have been taking place in midsummer 1972, although the sampling may have been insufficiently dense to adequately portray the situation.

During October 1962 the location of the Bering Sea Water core in the central Chukchi was farther north, and there was no evidence of residual cold water as far south as the latitude of Cape Lisburne. The position of the warm core at that time is inferred from Fig. 78, in which the temperature of water in the salinity band 32.3 to 32.9°/$_{00}$ is contoured. The salinity band used to represent the Bering Sea Water was chosen by examining stations occupied in the vicinity of Bering Strait; the highest salinity (32.9°/$_{00}$) was about 0.1°/$_{00}$ less than in the summer data, probably due to the small seasonal freshening trend noted previously. Though no indigenous Chukchi Sea water was observed at the latitude of Cape Lisburne, there were no stations for some distance to the north, between the warm

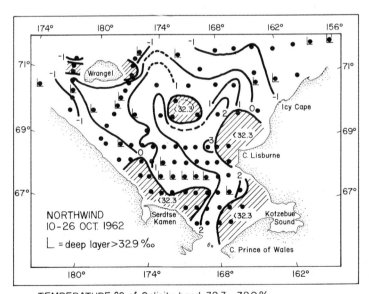

TEMPERATURE °C of Salinity band 32.3 – 32.9 ‰

78. Temperature of water with salinities between 32.3 and 32.9°/$_{00}$, 10–26 October 1962 (*Northwind*). "L" denotes a deeper layer with S > 32.9°/$_{00}$

core and Herald Shoal, so that we do not know whether the Bering Sea Water is always located well south of Herald Shoal or whether on occasion it fills the whole central region north to Herald Shoal.

With the location of the core of Bering Sea Water farther north in October 1962 than in August 1972, the front along the south and west sides was also displaced. In the south, it was farther offshore from Cape Serdze-Kamen, suggesting that greater amounts of Siberian Coastal Water had penetrated to the southeast. In the west, the front lay in Long Strait rather than east of Wrangel Island, suggesting that a net westward flow had occurred through Long Strait, including some Bering Sea Water (cf. section on *Currents* below).

The bulk of the Bering Sea Water turns northward at ~175°W longitude (Figs. 73, 78). In its subsequent progress toward Herald Canyon, it takes on more nearly the character of a narrow "river in the sea," closely bounded on the west by cold waters from the East Siberian Sea and, north of Herald Shoal, on the east by cold Chukchi Sea water. The nature of the water mass modification can be followed in the 1972 *Oshoro Maru* stations, for which the T-S relations are presented in Fig. 79.

The envelopes show progressive changes along the core, from where it turned north (stations 109 to 111) to Herald Island (stations 125 to 128). The water columns maintain a slight vertical salinity stratification, and considerable temperature stratification, but continue to show progressive cooling as during transit of the central Chukchi. However, it appears that vertical heat loss has increased significantly in this region, compared to farther south. From southwest of Herald Shoal to Herald Island, ~300 to 350 km, the minimum core temperature decreased more than 1°C or about 3×10^{-3} °C km^{-1}, which contrasts with a change throughout the central Chukchi of only $~1 \times 10^{-3}$ °C km^{-1}. Also in contrast with the changes in the central Chukchi, there is a noticeable vertical salt flux, with a progressive trend toward higher values of deep salinity. The mixing in this northwestern region is somewhat cross-isopycnal rather than along isopycnal surfaces, so that the water mass is in general becoming slightly denser as it progresses northward.

Also included in Fig. 79 are stations adjacent to the Bering Sea Water. Stations 115 and 124 border the Bering Sea Water on the west and station 129 on the east, while station 122 lies farthest north but close to the core. Station 115 shows no sign of Bering Sea Water, but demonstrates that the cold, flanking water has a broad range of salinities and densities. We interpret these features as follows. As the stream of Bering Sea Water narrows, it tends to interleaf with the cold, flanking waters. Mixing is predominantly along isopycnal surfaces, so that a layer of maximum temperature is created in the σ_t-range of the Bering Sea Water. The cold water

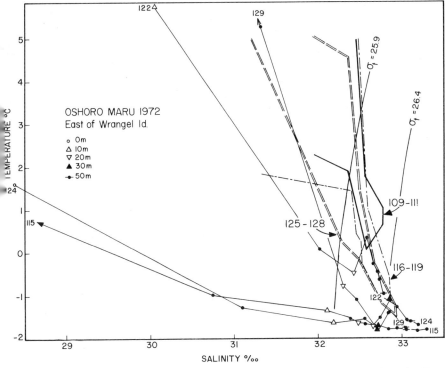

79. T-S diagram of 1972 *Oshoro Maru* stations east of Wrangel Island, and T-S envelopes from stations farther south (see Fig. 66)

of lower salinity, which is in large supply, tends to isolate the warmer more saline fraction leading to T-S curves with a subsurface temperature maximum as at station 122. The deeper Bering Sea Water tends to spread laterally over the cold, saline bottom water. The stations bordering the warm core, now increased about 0.2°/₀₀ in salinity near Herald Island (124 and 129), show the presence of the layering along $\sigma_t = 26.4$.

To summarize, Bering Sea Water entering the Chukchi forms the western part of the northward flow. The flow tends to follow the bottom contours, and so to some extent trends eastward into Kotzebue Sound, but by the latitude of Pt. Hope it is heading northwestward across the central Chukchi. There it interacts with a limited supply of a residual water from the previous winter. The remnants of this residual water are present in summer because they apparently lie in a dead spot, a cyclonic gyre, in the circulation, and therefore are not advectively removed. The Bering Sea Water is also bordered on the north by cold Chukchi water, but with which there appears to be little interaction. This water is probably also a residue from winter, but is being circulated more actively than the resident water to the south; e.g., it may be advected from north of Herald Shoal. There is some variability in the location of the Bering Sea Water and the cyclonic gyre, and probably also in the character of the water immediately south of Herald Shoal. At approximately 70°N, 175°W the Bering Sea Water turns northward and flows off the shelf over Herald Canyon.

The modification of this water mass through the central Chukchi Sea is not great. The core shows only a small cooling effect due to the interaction with limited amounts of cold residual water of approximately the same density. As resident water lies to the south, the cooling is progressively greater toward the south. Boundaries between Bering Sea Water and water on the south and west from the East Siberian Sea are characterized by water columns exhibiting a marked interleafing according to density.

On the northward leg the Bering Sea Water is flanked on both sides by cold water of both higher and lower salinity. Major modification takes place preferentially within the isopycnal band originally defining the Bering water, and the layering and mixing give rise to a subsurface temperature maximum. The vertical fluxes of heat and salt are now greater than in the central Chukchi, so that temperatures are depressed more and the deeper layers show a small increase in salinity. In the surface layers, freshening and cooling due to ice melt will further isolate the Bering Sea Water. This outflow of Bering Sea Water to the Arctic Ocean will in general occupy the σ_t band 26.0 to 26.5.

Alaskan Coastal Water

Alaskan Coastal Water is the low-salinity fraction of water entering the Chukchi from the Bering Sea. It can have a broad range of salinity, but has a relatively sharply defined upper limit (typically somewhat greater than 32°/$_{00}$) which is variable annually. For example, in summer 1967 the limit was about 32.4°/$_{00}$ (Figs. 67 and 68), in 1968 32.5°/$_{00}$ (Fig. 69), and in 1972 32.1°/$_{00}$ (Fig. 7).

Alaskan Coastal Water is the eastern part of the northward flow, and is joined by effluent from Kotzebue Sound. From the heat and salt content there is no way of distinguishing water from the two sources because they have similar T-S characteristics. However, there is usually some modification of the T-S limits of Alaskan Coastal Water by the time it reaches Pt. Hope. On 16 to 17 July 1967 (Fig. 68) it appeared to have about the same salinity limits as in Bering Strait, but the minimum temperature was reduced about 0.3°C. On 22 to 23 July (Fig. 70), on the other hand, it appeared that the salinity range had broadened slightly (the upper limit raised by 0.1°/$_{00}$ and the lower limit reduced by ~0.3°/$_{00}$), while minimum temperatures were nearly 2°C warmer. However, this latter modification may have been associated with the abnormal reversed flow conditions noted previously, which resulted in large amounts of Alaskan Coastal Water being backed up in Kotzebue Sound. In the July 1972 *Oshoro Maru* Cape Lisburne section (Fig. 77), which lies about 50 km farther downstream than the Pt. Hope section, the Alaskan Coastal Water had a broader salinity range than in Bering Strait. The upper limit of salinity had increased about 0.1°/$_{00}$, and the lower limit of deep salinities appeared to have decreased to <31°/$_{00}$. Minimum temperatures were about 0.2°C lower than in Bering Strait.

Thus Alaskan Coastal Water maintains its identity through the southern Chukchi Sea and remains distinct from the more saline Bering Sea Water, but its properties can undergo minor modification. This includes (1) a slight increase in the salinity (0.1 to 0.2°/$_{00}$) defining the upper limit of the range, probably due to some lateral interaction with the lowest salinity Bering Sea Water adjacent on the west, and (2) a decrease in the lower limit salinity and a cooling of the deeper layers in the lower part of the salinity range, probably due to admixtures of Kotzebue Sound water. The salinity range will vary from year to year, but in general is <31.0°/$_{00}$ to ~32.5°/$_{00}$. The minimum temperatures, which occur in the more saline water, are 1 to 3°C.

North of Pt. Hope-Cape Lisburne, the Alaskan Coastal Water turns northeast. Data in the region from Pt. Hope westward to Herald Shoal and northward to Icy Cape from the midsummer 1960 *Brown Bear* cruise are presented in Fig. 80 (for station locations see Fig. 66). All stations in the vicinity of Pt. Hope-Cape Lisburne (54 to 61) are enclosed by an envelope, which we take to represent the Alaskan Coastal Water during early August 1960. That year the low-salinity water with colder temperatures, which appeared to broaden the salinity range of the Alaskan Coastal Water in 1972 (Fig. 77), was not apparent; and, therefore, the T-S envelope has a long and narrow vertical aspect, subtending only 0.5°/$_{00}$ at temperatures <6°C. Though *Brown Bear* stations were too sparse near Bering Strait to clearly define the total T-S range of Alaskan Coastal Water, there had apparently been no great change in characteristics during the transit to Pt. Hope.

To the north of Cape Lisburne the Alaskan Coastal Water had turned northeast. The envelope of stations 67 to 70, which lay to the northwest, shows no water of Bering Sea origin. The core of Alaskan Coastal Water, in the σ_t band 25.0 to 25.8 in these data, was evidenced at stations 63 to 66 and 77. Its minimum temperature had, however, been reduced about 1°C in 100 km of travel. (Station 76 was located inshore from the core, in the

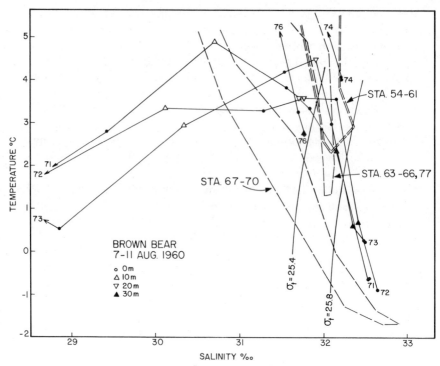

80. T-S diagrams and envelopes from northeastern Chukchi Sea, 1960 *Brown Bear*

gyre established in the lee of the peninsula, which is discussed in more detail below.)

The northernmost *Brown Bear* stations (71 to 73), off Icy Cape, show the Alaskan Coastal Water to spread above colder, more saline water. We interpret this to represent the confluence of two streams, the water from north of Herald Shoal merging with Alaskan Coastal Water. The former has the characteristics of residual Chukchi Sea winter water, being cold and relatively saline. In the vicinity of ice, local melting and cooling will sharply reduce the temperatures of near surface water, causing a hook down to the left in the T-S curve. This effect can be seen in the three northernmost stations (71 to 73), which were made close to the pack, station 72 actually being among drifting pans.

Vertical fluxes of both heat and salt cause the bottom water to be reduced in salinity and warmed, while the Alaskan Coastal Water increases in salinity as it is cooled. This water mass thus increases in density. The 1972 *Oshoro Maru* data demonstrate this change graphically. In Fig. 81 the envelope of stations 98 to 102 defines the Alaskan Coastal

Water off Cape Lisburne in late July 1972, and the arrow shows the direction in the T-S plane of the shift toward higher salinity and lower temperature in later stages of the water mass history (stations 133 to 139, Fig. 66). This outflow of water to the Arctic Ocean from the Bering Sea will thus occupy the σ_t band 25.5 to 26.0. West of the core of Alaskan Coastal Water, at stations 130 to 132, the densest water which was formerly underlying the core further south, has moved laterally, carrying with it its slightly higher temperatures and creating a slight deep temperature maximum. This appears to be analogous to the behaviour of the deep Bering Sea Water to the north of Herald Island.

The Alaskan Coastal Water is characterized by lateral gradation from a relatively cold and saline fraction on the west, to a warm and less saline one close to the coast. Figure 71 shows the warmest waters of the upper layer to extend from approximately the core of Alaskan Coastal Water eastward to the coast. The amount of heat transmitted downward, and the degree of warming of the underlying saline water, seems to be proportional to the temperature of the overlying water, so that stations

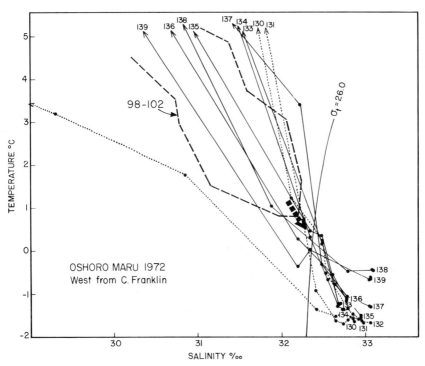

81. T-S diagrams and envelopes from northeastern Chukchi Sea, 1972 *Oshoro Maru* (Arrow indicates direction of change of T-S characteristic.)

138 and 139 (closest to the coast) show the highest deep temperatures. This may, of course, in part simply reflect this water having been exposed the longest to vertical heat flux.

The detailed data obtained north of Cape Lisburne and west of Icy Cape during *Websec-70* (Ingham et al. 1972) substantiate the divergence of the main stream of Alaskan Coastal Water from the coast north of Cape Lisburne and its subsequent return to the coast near Icy Cape. Figure 82 shows the *Websec-70* stations (25 September to 17 October 1970), where "A" denotes the presence of water with characteristics to the right of, on the T-S plane, a line through the points 32.2°/₀₀, 1°C and 31.5°/₀₀, 4°C (cf. Fig. 80). Water more saline and warmer than this identifies the deep core of Alaskan Coastal Water, while that less saline and cooler is surface water. The measurements suggest an anticyclonic gyre inshore of the Alaskan Coastal Water flow, in the lee of the Pt. Hope-Cape Lisburne peninsula; this can also be inferred from the 1960 current measurements (Fig. 46). As the 1970 data were taken during late September/October and some local freezing was occurring, surface temperatures in the inshore part of the gyre had all been reduced to <0°C, and the T-S curves showed the inverted U shape typical of fall.

The modification of the Alaskan Coastal Water during its northeast-

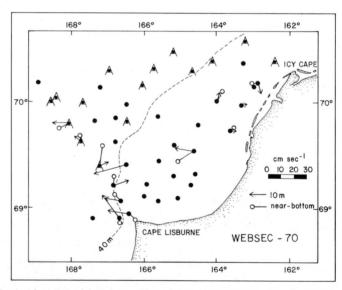

82. Station locations of *Websec–70* (Ingham et al. 1972) (Vectors are of 0 to 10 m and near-bottom currents averaged over various intervals between 2.5 and 31 hours. "A" denotes water in the column more saline and warmer than a limiting condition of 32.2°/₀₀, 1°C and 31.5°/₀₀, 4°C.)

ward travel appears similar to that of the Bering Sea Water on its northward transit past Herald Island. An estimate was made of the vertical eddy conductivity in both of these water masses as follows. The changes in average temperature of the 10 to 30 m layer for groups of *Oshoro Maru* stations along the water mass cores were assumed to be due to the net heat gain from the 0 to 10 m layer minus that lost to the layer deeper than 30 m:

$$\frac{1}{A} \frac{\Delta Q}{\Delta t} = \rho C_P K_T \left[\frac{\Delta T}{\Delta z}\bigg|_U - \frac{\Delta T}{\Delta z}\bigg|_L \right]$$

where A is the area, $\Delta Q/\Delta t$ is the rate of change of heat content in 10 to 30 m layer (Δt was taken as the appropriate station separation divided by the mean measured flow speed), $\rho C_P = 1$ cal cm^{-3} °C^{-1}, K_T is the eddy conductivity, and $\Delta T/\Delta z$ the temperature gradient between the 0 to 10 m and 10 to 30 m layers ($_U$) and 10 to 30 m and >30 m layers ($_L$), respectively. Using upstream station groups 98 to 102 for the Alaskan Coastal Water and 109 to 111 for the Bering Sea Water, and stations 136 to 139 and 116 to 119 for the corresponding downstream groups, the calculations gave $K_T = 0.2$ cm^2 sec^{-1} and 2.5 cm^2 sec^{-1}, respectively.

Finally we note that in the Arctic Ocean water from the Bering Sea has been identified by Coachman and Barnes (1961) by a slight temperature maximum in the upper part of the pycnocline, centered at ~75 m depth. The T-S diagram basic to their analysis is shown in Fig. 83. We can now conclude that the water identified in the Canadian Basin of the Arctic Ocean was contributed solely by the northeast, or Alaskan Coastal Water, branch of the Bering Sea input. The σ_t band 25.5 to slightly over 26.0,

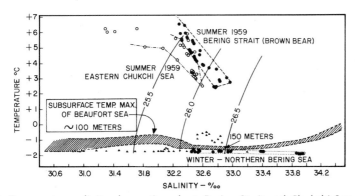

83. Temperature-salinity observations from Bering Strait and Chukchi Sea compared with the Arctic Ocean (Coachman and Barnes 1961, Fig. 6, to which has been added the isopycnal $\sigma_t = 26.5$). (Triangles denote observations from 1922 *Maud* near Herald Island.)

delineates the northeast branch water; while northwest branch water, in the σ_t band 26.0 to 26.5, does not appear in these Arctic Ocean data. However the *Maud* data in Fig. 83 clearly show water of the northwest branch occurring over Herald Canyon from Herald Island northward. Certainly northwest branch water contributes to the Arctic Ocean subsurface layer, and there are two possible reasons why its temperature maximum has not been identified in the Arctic Ocean: (1) It is >200 km from Herald Island to the continental slope (contrasted with the ~150 km from Pt. Franklin into the Canadian Basin), so that vertical heat flux may be sufficiently large to eliminate any distinguishable temperature maximum (in this connection we note that the calculated eddy conductivity for the northwest branch was one order of magnitude larger than for the northeast branch, though such estimates are admittedly crude); and/or (2) the northwest branch enters the Canadian Basin to the west of the Chukchi rise, while the northeast branch enters directly into the Beaufort gyre, and since the data used to identify Bering Sea water were all from the Beaufort gyre, its presence may simply have been overlooked.

CURRENTS

Current measurements from the Chukchi Sea are few, and most available data have already been presented (Figs. 46, 47). The most comprehensive measurements are those from *Oshoro Maru* in July 1972, from which the average flow in the upper (0 to 10 m), intermediate (10 to 30 m), and bottom (>30 m) layers are shown in Fig. 84. Comparison of Figs. 46, 47, and 84 indicates a relatively consistent flow pattern which agrees with that deduced from the water mass analysis. We show schematic interpretations of the mean flow in the upper and lower layers in Figs. 85 and 86. Also included in Fig. 85 are various positions to be expected for the front between Bering Sea Water and water derived from the East Siberian Sea, as well as estimates of representative flow speeds in various locations. In Fig. 86 are indicated approximate positions of the cores of Bering Sea and Alaskan Coastal waters.

The swiftest flows are always encountered in the eastern channel of Bering Strait, where speeds in excess of 150 cm sec^{-1} are common. In the western channel speeds are normally ~30 cm sec^{-1}. One hundred kilometers north from the strait the swiftest speeds (~50 cm sec^{-1}) are over the relatively deep channel to the west of Cape Prince of Wales Shoal. Outer Kotzebue Sound, where the bathymetry and the flow trends east and then north, is an area of marked deceleration, with speeds of 15 to 20 cm sec^{-1}.

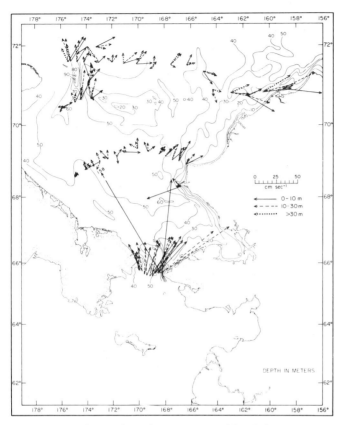

84. Currents averaged over three layers measured by *Oshoro Maru*, 24 July–1 August 1972

General acceleration of the flow (to ~40 cm sec^{-1}) occurs near the shore region along the south side of the Pt. Hope-Cape Lisburne peninsula. Otherwise, over the whole south-central Chukchi speeds of 15 to 25 cm sec^{-1} are indicated. A small acceleration (to ~30 cm sec^{-1}) seems to occur where the northeast branch closes the Alaskan coast near Icy Cape. No significant acceleration of the flow appears to occur over Herald Canyon, where speeds of 15 to 25 cm sec^{-1} to the north were also observed from the *Maud* in 1922 (Sverdrup 1929). No data are available to provide speed estimates for the Siberian coastal flow, but speeds of 10 to 30 cm sec^{-1} have been observed in Long Strait (see below).

The *Oshoro Maru* data have been used to estimate the transport of the water masses, and the calculations are summarized in Table 13. We cannot anticipate a completely balanced mass budget, because the sections

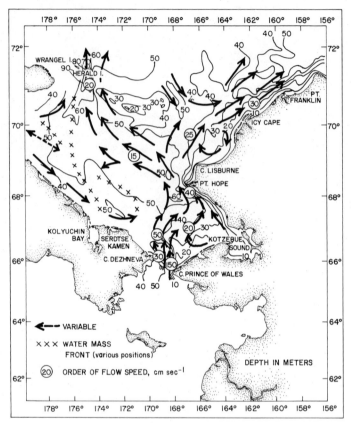

85. Schematic of upper layer flow in the Chukchi Sea. (Dotted arrows indicate variable currents. Various positions of water mass fronts are indicated, and circled numbers are estimated flow speeds in cm sec^{-1}.)

were occupied sequentially over the period 24 July to 1 August and, as has been observed, transports through the system can vary by a few tenths of a Sverdrup in 1 to 2 days. However, the calculations do suggest some generalizations about the quantitative aspects of the passage of water from the Bering Sea through the Chukchi.

The total transports suggest that the system slowed down between the times of occupation of the Bering Strait section and the Cape Lisburne section (cf. Fig. 62b), but may have resumed a greater northward flow by 31 July to 1 August. However, the net northward transport of 2.3 Sv through the last section was sufficiently greater than a typical Bering Strait transport value to suggest that considerable (perhaps one-half Sv) resident Chukchi and East Siberian Sea water was entrained with the northward flux.

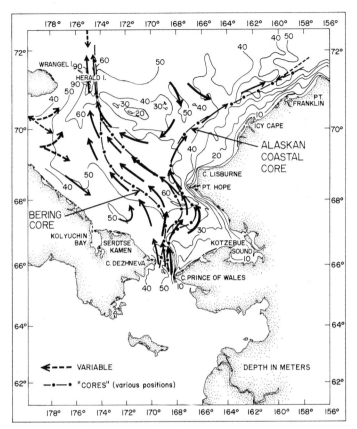

86. Schematic of lower layer flow in the Chukchi Sea. (Dotted arrows indicate variable currents. Various positions of "cores" of Bering Sea water mass are indicated.)

TABLE 13
Transport (Sv; + north) of Water Masses, July-August 1972
[Oshoro Maru; Station Nos. in ( ) from Figs. 7, 79, 81, 83]

Section	Date	Bering Sea	Alaskan Coastal	Section Total
Bering Strait	7/24–25	1.1 (89–97)	0.6 (85–88)	+1.7
Lisburne	7/27–28	0.2 (109–111)	0.7 (98–102)	+1.3
SE of Wrangel	7/29	0.8 (116–119)		
Herald Island-C. Franklin	7/31–8/1	0.3 (125–128)	1.3 (133–139)	+2.3

The transport variations of waters that can be positively identified as belonging to each of the two water masses, Bering Sea and Alaskan Coastal, appeared to differ. Water identifiable as Alaskan Coastal seemed to increase in volume, presumably due to additions from Kotzebue Sound and possibly from entrained Chukchi Sea water north of Cape Lisburne. Identification of the latter Alaskan Coastal Water would have to be caused by sufficient additions of heat to the more saline fraction, such that temperatures were not reduced much below 0°C (cf. Fig. 81). In contrast, the transport of identifiable Bering Sea Water appeared to decrease northward, while the total transport was increasing due to entrainment. The implication is of a greater net mixing and exchange of properties of the Bering Sea Water, both with the southern resident water while crossing the central Chukchi Sea and with surrounding waters on the northward leg toward Herald Island, than is the case with the Alaskan Coastal Water. The latter is apparently somewhat isolated, perhaps in part by the physical boundary of the Alaskan coast, while the Bering Sea Water passes through a more open oceanic environment. In this connection we note the similar conclusion of the previous section, based on other lines of evidence.

There are few data to assess the time-variability of the currents. The only moored current meter records are from immediately north of Bering Strait, and they were discussed in Chap. 3. We can judge from the water mass distributions that considerable variations are superimposed on the mean flow pattern. For example, Aagaard (1964) reported that during October 1962 the surface flow of Alaskan Coastal Water toward the northeast had apparently stopped. We have recently analyzed long-term current meter records from Barrow Canyon to determine the time-dependence of that flow and these results are given in detail elsewhere (Mountain et al., in press).

During 12 to 24 August 1966 eleven anchored current stations were occupied in Long Strait (Coachman and Rankin 1968) which provide data on the magnitude and variability of the flow between the East Siberian and Chukchi seas. The results suggest that there is no major long-term net exchange between these seas, but rather that the net flow is east or west depending on the regional atmospheric conditions and continuity requirements. When winds were from the north over the Chukchi or from the south over the East Siberian Sea, and predominantly easterly in Long Strait, the net water motion in Long Strait was to the west, and vice versa. As an index to the regional winds, negative values of the surface atmospheric pressure differences (Pt. Barrow − Wrangel Island) and (Kolyma River delta − Wrangel Island) would be associated with the former condition, and vice versa. This agrees with the conclusion of Gorbunov (1957),

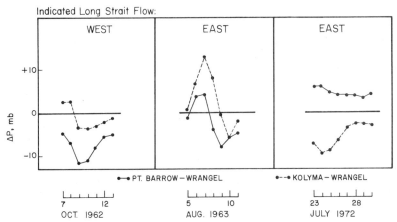

87. Three-point running means of surface daily (1200 Z) atmospheric pressure differences to the east (Pt. Barrow) and west (Kolyma River) of Wrangel Island for three periods, when oceanographic data indicate east or west flow through Long Strait

who correlated the drift of a radio buoy in Long Strait with the predominance of easterly winds.

There are three periods when the water mass data suggest that there must have been net flow through Long Strait. During the period 10 to 12 October 1962 (Fig. 78) Bering Sea Water was observed south and southwest of Wrangel Island, so that at least in the north portion of the strait there must have been westerly flow. Conversely, during the periods 8 to 10 August 1963 and 27 to 29 July 1972 it appears that the Bering Sea Water had turned sharply north well to the east of Long Strait and that there was a net flow to the east of water from the East Siberian Sea. Figure 87 correlates these deduced flows with the atmospheric pressure differences cited above. The results are suggestive though not conclusive. The westward intrusion of Bering Sea Water through Long Strait during October 1962 coincides with a period when the atmospheric pressure differences would indicate both a northerly wind over the Chukchi and a southerly wind in the East Siberian Sea. The easterly flow in August 1963 immediately follows a period where the wind regime is opposite, while in July 1972 it occurred during nearly neutral wind conditions (southerly everywhere).

5

NUMERICAL CONSIDERATIONS

CERTAIN observational programs recently conducted in the region, notably those undertaken during the summers of 1967 (*Northwind*) and 1968 (*Staten Island* and *Thompson*), were designed to provide data suitable to various calculations which could give a more complete description of the oceanographic regime. These include dynamical calculations and estimates of eddy coefficients and heat fluxes, which are considered in this chapter.

DYNAMICAL CALCULATIONS

In the left-hand coordinate system with x in the direction of flow (north) and z positive downward, the x-component equation of motion is

$$\frac{du}{dt} = -\frac{1}{\rho}\frac{\partial p}{\partial x} + 2\omega \sin\phi\, v + \frac{1}{\rho}\left\{\frac{\partial \tau_{xx}}{\partial x} + \frac{\partial \tau_{xy}}{\partial y} + \frac{\partial \tau_{xz}}{\partial z}\right\} \quad (1)$$

in which u and v are, respectively, the x- and y-components of velocity; ρ is density; p is pressure; $2\omega \sin\phi$ is the Coriolis parameter; and τ_{xx}, τ_{xy}, and τ_{xz} are the tensor components of stress in the x direction. Coachman and Aagaard (1966) noted that in application to Bering Strait $\partial \tau_{xx}/\partial x$ and $\partial \tau_{xy}/\partial y$ had to be neglected as a practical matter. With x taken north in the strait (the flow direction) and the individual acceleration term replaced by local and inertial terms, (1) then becomes

$$\frac{\partial u}{\partial t} + u\frac{\partial u}{\partial x} + w\frac{\partial u}{\partial z} = -\frac{1}{\rho}\frac{\partial p}{\partial x} + \frac{1}{\rho}\frac{\partial \tau_{xz}}{\partial z} \quad (2)$$

Assuming hydrostatic equilibrium and water to be incompressible, (2) can be integrated from the free surface $-\zeta$, the elevation above undisturbed sea level $z = 0$, to the bottom $z = h$, giving

$$\int_{-\zeta}^{h} \frac{\partial u}{\partial t} \, dz + \frac{\partial}{\partial x} \int_{-\zeta}^{h} u^2 dz + \int_{-\zeta}^{h} u \frac{\partial v}{\partial y} \, dz - \frac{1}{\rho}(\tau_h - \tau_o)$$

$$+ \frac{g}{\rho} \int_{-\zeta}^{h} \left[\int_{-\zeta}^{z} \frac{\partial \rho}{\partial x} dz \right] dz + gh \left[\frac{\partial \zeta}{\partial x} - \frac{\partial P_o}{\partial x} \right] = 0 \quad (3)$$

where g is the acceleration due to gravity, ρ is the mean density of the water column, τ_o and τ_h the shear stress at the free surface and the bottom, respectively, and P_o the atmospheric pressure (in cm of water) on the free surface (cf. Coachman and Aagaard 1966). With certain assumptions, the first five terms can be estimated from the data for the various sections and the sixth from atmospheric data, so that the longitudinal barotropic pressure gradient term, $gh(\partial \zeta/\partial x)$, associated with the slope of the sea surface, can be determined by difference.

The results of these dynamical calculations are brought together in Table 14. In addition to the results for the 1964 *Northwind* section, reproduced here from Coachman and Aagaard (1966) as averages of the sets of values reported earlier, the other suitable data were from 1967 *Northwind*, 1967 *Thompson*, 1968 *Staten Island*, and 1973 *Thompson*. Assumptions and manipulations of the data followed as nearly as possible the procedures of Coachman and Aagaard, but did differ in certain cases:

(1) Local acceleration $\left(\int_{-\zeta}^{h} \frac{\partial u}{\partial t} dz \right)$: For the 1967 *Northwind* sections, the x-component of velocity averaged across the strait layer by layer was compared between the first section (13 to 14 July, B in Fig. 2a) and the second (18 to 19 July, B in Fig. 2b), and likewise for the 1973 *Thompson* (29 to 30 Sept. with 3 to 4 Oct.). For the calculations north of the strait in 1967, the repeated crossings of the Shishmaref section (C in Figs. 2a, b) were used. For the 1967 *Thompson* data the changes were calculated between each four section repetitions. For the 1968 *Staten Island* data, the section south of the strait (C, Fig. 3) was extrapolated into the strait by increasing the speeds proportional to the layer-by-layer cross-sectional area reduction for each water mass, and the resulting speeds compared with those of section D in the strait. We note that this procedure resulted in a fairly large value for local acceleration ($+3.5$ cm^2 sec^{-2}). However, the system *was* accelerating markedly at the time, as witnessed by the increase in transport from section D to section E of 0.7

TABLE 14
Dynamical Calculations for Bering Strait

Parameters [eqn. (3)]:

$$A \quad \int_{-\zeta}^{h} \frac{\partial u}{\partial t} dz \qquad C \quad \int_{-\zeta}^{h} u \frac{\partial v}{\partial y} dz \qquad E \quad \frac{g}{\rho}\left(\int_{-\zeta}^{h}\left[\int_{-\zeta}^{z} \frac{\partial \rho}{\partial x} dz\right] dz\right) \qquad G \quad gh \frac{\partial \zeta}{\partial x}$$

$$B \quad \frac{\partial}{\partial x}\int_{-\zeta}^{h} u^2 dz \qquad D \quad -\frac{1}{\rho}(\tau_h - \tau_o) \qquad F \quad -gh\frac{\partial P_o}{\partial x}$$

Results:

cm² sec⁻²; + north

Case	A	B	C	D	E	F	G	Transport Sv.	Slope × 10⁶
1. August 1964	−1.0	−0.7	−3.2	−3.9	−1.6	−0.3	+10.7	+1.4	+2.2
2. July 1967	+0.4	−0	−0.1	+2.4	+0.3	+0.2	−3.2	−0.2	−0.7
3. July 1967 (north of strait)	−0	−0	−0.2	−0.3	−1.0	+0.4	+1.1	0	+0.2
4. August 1967 (1–2)	−1.5	−0.7*	−2.2	−9.4	+0.5	−0.4	+13.7	+1.6	+2.8
5. (2–3)	+0.3	−0.7*	−2.2	−7.3	−1.2	−0.6	+11.7	+2.0	+2.4
6. (3–4)	−1.0	−0.7*	−2.4	−6.7	+1.1	−0.8	+10.5	+1.4	+2.1
7. July 1968	+3.5	+2.5	−7.0	−6.4	−0.7	+0.4	+7.7	+1.5	+1.6
8. October 1973	−0.1	−0.9	−0.5	−1.6	−2.3	+0.2	+5.2	+1.4	+1.1
Mean Values	+0.1	−0.2	−2.2	−4.2	−0.6	−0.1	+7.2		

*1964 value, see text.

Sv in one day (Fig. 61b), so that the result does not appear to be completely unreasonable.

(2) First inertial term $\left(\dfrac{\partial}{\partial x} \displaystyle\int_{-\zeta}^{h} u^2 dz \right)$: For the 1967 *Northwind*, 1968 *Staten Island*, and 1973 *Thompson* data, the section immediately upstream (south) was used for the longitudinal comparison. There appeared to be no reasonable way to calculate this term for the 1967 *Thompson* sections, and so the value from the 1967 *Northwind* data was used, the rationale being that transports and accelerations were nearly the same for these data. Furthermore, this inertial term does not appear to be particularly significant at such times and should not seriously affect the calculation of $gh(\partial \zeta/\partial x)$.

(3) Second inertial term $\left(\displaystyle\int_{-\zeta}^{h} u \dfrac{\partial v}{\partial y} dz \right)$ and net stress on the water column $\left[-\dfrac{1}{\rho}(\tau_h - \tau_o) \right]$: These were calculated from the section data according to the methods of Coachman and Aagaard.

(4) Longitudinal baroclinic pressure gradient $\left(\dfrac{g}{\rho} \displaystyle\int_{-\zeta}^{h} \left[\displaystyle\int_{-\zeta}^{z} \dfrac{\partial \rho}{\partial x} dz \right] dz \right)$:

For July 1967, both in and north of the strait, July 1968, and October 1973 this term was evaluated by comparison between the upstream section and the section under consideration. For the 1967 *Thompson* cruise, the data from the second crossing (5 August) were extrapolated upstream station by station for a distance $\overline{U} \Delta t$, where $\overline{U}$ was the mean local speed of the north component of flow and Δt the time interval between the first crossing on 4 August and the second crossing. The same procedure was applied to the second and third, and third and fourth crossings.

(5) Longitudinal pressure gradient on the free surface [$-gh(\partial P_0/\partial x)$]: Values of surface atmospheric pressure for Cape Lisburne (68°53′N) and Northeast Cape, St. Lawrence (63°19′N) were obtained from 1200 Z synoptic surface pressure charts for the day of section occupation and the day prior, averaged, and the gradient computed by dividing the difference (in cm of water) by 618 km.

The results (Table 14) substantiate the earlier conclusion of Coachman and Aagaard (1966), viz., that even though the mean flow through Bering Strait is approximately antitriptic (longitudinal pressure gradient balanced by shear stress), other parameters must also be retained to portray the flow regime adequately. By far the most important of these is the second inertial term (C), representing the lateral advection of momentum. Together with the shear stress term, it in the mean balances the barotropic component of the pressure gradient to within 11%. That nonlinear effects should be important is not surprising, for the Rossby number ($R_o = (\partial u/\partial y) / f$ = relative vorticity/planetary vorticity) in Bering Strait is quite large, being on the order of 0.5. It appears that on the whole the secondary forces all tend to oppose the flow through the strait being driven by the surface slope. It is notable, however, that in the mean the baroclinicity is negligibly small in the total momentum balance, although it may in any particular instance be significant. The implication is either that on the whole baroclinic adjustment is decoupled from the surface slope, or that the density field is sufficiently noisy to make it appear random. We note that the former condition was inferred independently in Chap. 3 from linear trends in moored current meter records.

The north-south atmospheric pressure gradient on the sea surface appears to provide only minor modification to the flow in the strait proper, as values for term F are the same order as those for baroclinicity (E) and longitudinal inertial acceleration (B). This does not negate the possibility that variations in the meridional atmospheric pressure gradient may cause major modifications at other locations in the regional flow field where the force balance is not so closely antitriptic, as suggested by the Barrow Canyon results.

We had anticipated previously (Coachman and Aagaard 1966) that away from Bering Strait the surface slope would be less. The present results can be interpreted as support because in the one case calculated (3) the surface slope in the direction of flow was quite small and we might expect the flow regime in this case to have been more nearly geostrophic.

That the transport through Bering Strait is governed by the sea surface slope seems indisputable. The correlation ($r = 0.92$) between transport and surface slope is shown in Fig. 88, together with the regression line. Thus, with the flow through Bering Strait predominantly to the north, sea level in the Chukchi Sea is predominantly lower than in the northern Bering Sea. Occasional flow reversals to the south occur only when sea level in the Chukchi is raised temporarily higher than in the northern

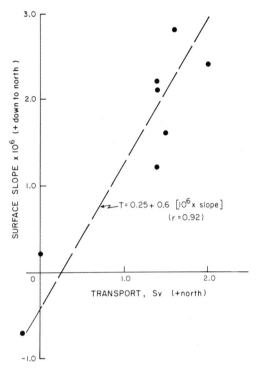

88. Correlation of surface slope in Bering Strait with transport

Bering (case 2). For an average summer transport of 1.7 Sv, the slope is 2.4×10^{-6}.

In addition to the regional wind regime maintaining the mean sea surface slope, it is conceivable that a slope of proper magnitude could also be sustained by the differences in steric level between the Pacific and Arctic oceans. Table 15 contrasts approximate mean dynamic height anomalies for the Aleutian Basin of the Bering Sea against those of the Canadian and Eurasian basins of the Arctic Ocean. If the integrated dynamic heights above 1500 db are representative of water column heights, the surface of the northern extremity of the Pacific would seem to be about 1 m higher than that of the Arctic, providing a slope of order 10^{-6} to 10^{-7} down to the north.

We feel that while this steric difference could provide the pressure head to drive the mean flow north through Bering Strait, no firm conclusion is possible at present. The actual transport at any time, though, is certainly significantly influenced by the regional atmospheric pressure (wind) re-

BERING STRAIT: The Regional Physical Oceanography

TABLE 15
Approximate Mean Dynamic Depths, 10/1500 db*

	ΔD, dyn m	Steric Height, m
Aleutian Basin, Bering Sea	1.50	1474.6
Canadian Basin, Arctic Ocean	0.60	1473.7
Eurasian Basin, Arctic Ocean	0.40	1473.5

*Averaged over a number of stations from each basin.

gime and its effects in moving water in the shallow and bounded Bering and Chukchi seas.

Shtokman's (1957) conclusion that purported seasonal fluctuations in the Bering Strait flow (winter transport being only one-fourth that of summer) are caused by seasonal steric level variations in the Bering Sea seems incorrect, for the maximum seasonal variation in dynamic heights that we can envision is only about 0.2 dyn m (based on a differential warming of 10°C and a freshening of 1°/$_{oo}$ in the upper 100 m).

EDDY COEFFICIENTS

Data from the 1968 *Staten Island* cruise, in which closed sections across the sea from St. Lawrence Island north through Bering Strait were occupied (Fig. 3), are suitable for estimation of vertical and horizontal eddy coefficients. Two methods were used. Basic to both methods is the assumption that salt and heat in the layer beneath the pycnocline (~10 m) were conserved.

Method 1

Under the above assumption, the individual rates of change of salinity and temperature of the lower layer are due solely to vertical and lateral eddy diffusion, i.e.,

$$\frac{dS}{dt} = K_H \frac{\partial^2 S}{\partial y^2} + K_V \frac{\partial^2 S}{\partial z^2} \qquad (4)$$

$$\frac{dT}{dt} = K_H \frac{\partial^2 T}{\partial y^2} + K_V \frac{\partial^2 T}{\partial z^2} \qquad (5)$$

where K_H and K_V are the (constant) eddy coefficients of the horizontal (H) and vertical (V) fluxes of salt and heat, S is the salinity, and T is the temperature. The mean values of salinity and temperature within each water mass in each section were determined and compared section by section from south to north using section separation and mean current speeds to estimate the individual rates of change. The curvatures of the salinity and temperature fields were assessed from the station data using the method of Proudman (1953). For lateral curvature, stations across the water mass boundaries were used; and for vertical curvature, average values within each water mass at 5, 10, and 15 m or at 10, 15, and 20 m, depending on pycnocline depth. The equations were then solved simultaneously for K_H and K_V.

Method 2

The core of each water mass (Fig. 12) can be identified by the minimum value of temperature of the deeper water in the salinity band defining the water mass. The gain in heat in the water column beneath the pycnocline must be due to the vertical flux down from the upper layer and the lateral flux from the adjacent water mass, i.e.,

$$L \frac{\Delta Q}{\Delta t} = C_P \left[K_H \frac{\Delta T}{\Delta y} + K_V \frac{\Delta T}{\Delta z} \right] \qquad (6)$$

where L = distance between sections, Q = heat, and C_P = specific heat. Similarly for salt conservation,

$$L \frac{\Delta S}{\Delta t} = K_H \frac{\Delta S}{\Delta y} + K_V \frac{\Delta S}{\Delta z} \qquad (7)$$

Two or three stations were averaged to define values for each core in each section; vertical gradients were estimated using the differences between mean upper-layer (0 to 10 m) and midlayer (10 to 30 m) values, and horizontal gradients on each side of the core calculated from the mean midlayer (10 to 30 m) values and the horizontal geometry of Fig. 12. The equations were then solved simultaneously for K_H and K_V.

The results of these calculations are shown in Table 16. There are numerous uncertainties in the calculations and hence we do not take the values too literally. Notably there is no way to suppress short-period variability in the system, and calculations involving salinity values are tenuous because numerical differences are almost always small. In the gross sense, though, we believe the results to be informative, and on the average to be reasonable.

There is the suggestion in Table 16 that lateral diffusion has an increasingly more significant effect eastward across the sea. Vertical fluxes across the pycnocline in summer are associated with eddy coefficients of order 1 cm^2 sec^{-1}. The values of K_V decrease with increasing stability through the pycnocline. In Fig. 27 average values of K_V from Table 16 have been plotted on the distribution of maximum stability in approxi-

TABLE 16
Eddy Coefficients from 1968 Staten Island

Section (Fig. 3)	Anadyr		Bering Shelf		Alaskan Coastal	
	(1)*	(2)*	(1)	(2)	(1)	(2)
	K_V cm² sec⁻¹					
Shishmaref (E)						
Bering Strait (D)	0.6	7	2	37	0.4	22
King Island (C)	1	5	0.7	45	1	4
C. Rodney (B)	4	}0.8	1	3	0.5	}0.3
St. Lawrence (A-A')	2		1	3	0.3	
C. Romanzof (A)			0.2	4		
	K_H 10⁵ cm² sec⁻¹					
Shishmaref (E)						
Bering Strait (D)	2	0.3	19	2	57	0.2
King Island (C)	0.4	0.1	4	2	48	0.6
C. Rodney (B)	2	}1	23	0.1	18	}0.1
St. Lawrence (A-A')	1		43	0.1	102	
C. Romanzof (A)			2	0.1		

*(1) Method 1, (2) Method 2 (see text).

mately their applicable positions. The low value (0.3 cm² sec⁻¹) of K_V along the Alaskan coast applies to a region of high stability; whereas north of Bering Strait, where the stability is quite low, K_V may be as high as order 10 cm² sec⁻¹.

HEAT BUDGET

The 1968 *Staten Island* series of sections also provide data from which the heat budget can be estimated. It was first assumed that each water mass was laterally isolated, i.e., that its heat content was changed only by vertical fluxes. (Thus, any net local budget imbalance, above and beyond what is attributable to surface flux, must in reality be due to lateral heat exchange.) The weighted (by area) average total heat content of the bottom layer (30 m to bottom) for each water mass and each section was calculated and compared with the next section downstream, so that by the above assumption any gain or loss is attributed to vertical exchange with the 10 to 30 m layer overlying. The heat balance of the 10 to 30 m layer was then in turn considered to have been satisfied by a flux from the surface (0 to 10 m) layer, and the balance for the surface layer achieved by exchange across the sea surface. The results of these calculations are plotted in Fig. 89.

Our first conclusion is that the overall mean surface flux is somewhat higher than can be accounted for by direct exchange at the sea surface. The weighted average value for the entire area (excluding the small area of very high values just SE of Bering Strait) is about 200 ly day^{-1}, whereas the net radiative flux above is only about 150 ly day^{-1} (Neumann and Pierson 1966, Fig. 9.9, p. 241), from which must also be subtracted the net evaporative minus sensible loss. We feel this discrepancy is not primarily due to noise in the calculations, but instead points toward the importance of Norton Sound (which was not included in the calculations) as a source of summer heat for the northern shelf.

Secondly, though we cannot take the individual values too literally, there appears to be a more or less systematic pattern to the surface heat

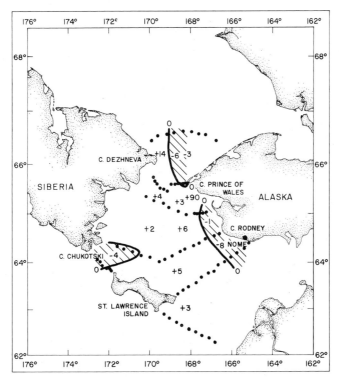

89. Heat budget for a column of water (10^2 ly day^{-1}), 9–19 July 1968

exchange. Over most of the region the values reflect the gain of heat to be expected for the location and time of year. However, there appear to be three areas of net heat loss even at a time when the surface waters of the region were absorbing significant amounts of heat. Two are associated with Alaskan Coastal Water, viz., along the Alaskan coast from Nome to Port Clarence, and along Cape Prince of Wales shoal. The third area is Anadyr Strait. (The high value for heat gain in Alaskan Coastal Water west of Port Clarence [SE of Cape Prince of Wales] must be attributed mostly to lateral heat exchange with the relatively shallow embayment to the east, where the water would locally have accumulated anomalously large amounts of heat.*) These deficits must be due primarily to a lateral loss of heat. In connection with the heat loss from Alaskan Coastal Water we note the presence in the eastern Bering Sea of the largest lateral eddy coefficients, referred to earlier. It is also notable that the largest local heat gain values (+ in Fig. 89) are those in areas immediately adjoining the regions of heat deficit, as would be expected if the former areas were receiving heat transferred laterally.

*Embayments isolated from main circulations have less effective heat removal; therefore surface heat additions (insolation, warmed runoff, etc.) are more effectively retained.

6

SUMMARY AND CONCLUSIONS

BERING STRAIT provides a water connection between two of the largest continental shelves of the World Ocean, the eastern Bering Sea shelf and the Siberian shelf, and it is the pathway for water exchange between the Pacific and Arctic oceans. The continental land masses of Alaska projecting west and Siberia projecting east are here separated by only 85 km, producing a funnel-shaped restriction on the dominant northward flow. During recent geologic time (~13,000 years B.P.) this great plain was subaerial with profound consequences for intracontinental migration of fauna, but at present it lies submerged to shallow, quite uniform depths of 50 m.

The water in shallow regimes, where a large amount of surface is exposed relative to the water volume, is strongly influenced by surface exchanges of heat, mass, and momentum. These exchanges condition and modify the properties of the shelf waters, leading to considerable diversity in characteristics, and in large measure control their circulation. Additionally, important effects in the Bering Strait region might be expected to be associated with the seasonal ice cover; the fresh water runoff, which is extremely variable seasonally and introduced primarily along the eastern (Alaska) side; and the regional wind-produced setup of sea level against the Siberian and Alaskan land masses.

Modern oceanographic exploration in the region began with Sverdrup's observations from the *Maud* in 1922. The major Soviet expeditions under Ratmanov in 1932–1933 and U.S. expeditions under Barnes and Thompson in 1933–1934 were the first systematic studies. Altogether some 25 cruises to the region, half of them conducted during the last decade, have produced significant data. In this book we have attempted to synthesize this material into a coherent description of the water masses and their origins, modifications, and movements, along with the flow field and its variability.

The presentation has been ordered on a subregional basis, with the hope that such organization will prove most useful to the reader. Water

158 • **BERING STRAIT: The Regional Physical Oceanography**

masses identifiable in Bering Strait were discussed first (Chap. 2), then traced upstream (south) to their origins in the Gulf of Anadyr and on the Bering Sea shelf. This was followed by discussion of temporal changes, including seasonal cycles and modifications, long-term trends, and the relationship to runoff and surface exchange. Chapter 3 described the flow field of the same subregion. The transport through Bering Strait was related to the regional atmospheric pressure, and time-dependent current variations and their causes were discussed. The Chukchi Sea was treated in Chap. 4, including the water masses and their modifications and variations, together with the flow field. The dynamics of the flow regime in Bering Strait were elucidated in Chap. 5, which also included calculations of eddy coefficients and a regional heat budget. The major results of the analyses are summarized below.

WATER MASSES

Three water masses are identifiable in Bering Strait, based on differences in salinity. The most saline water mass (Anadyr) lies on the west side, the least saline (Alaskan Coastal) along the east, and between lies Bering Shelf Water. There is very little lateral mixing apparent in the region north of St. Lawrence Island to Bering Strait and, therefore, salinity is largely conservative and the water masses remain easily distinguishable. On the other hand, heat within a water column is not conserved and therefore temperature is not a useful identifying characteristic. Anadyr Water attains its characteristics (S $\approx$ 32.8 to 33.2$°/_{00}$) in the Gulf of Anadyr, but the water is not endemic to the Gulf. Rather, water from the Bering Sea to the south enters and moves around the gulf on its way to Bering Strait, largely steered by the bathymetry, and this relatively warm (1 to 2°C) and saline ($>33°/_{00}$) water mixes with a cold, less saline central Gulf of Anadyr water; the resulting Anadyr Water is about 80 to 90% of Bering Sea origin and 20 to 10% resident Gulf of Anadyr water. Bering Shelf Water enters the northern Bering Sea from around both ends of St. Lawrence Island. It is formed from a portion of the Bering Sea water entering the Gulf of Anadyr from the south which undergoes significant lateral mixing and interspersal with cold, resident shelf water in the region south of St. Lawrence Island. The product is a water mass of distinctly lower salinity (S $\approx$32.4 to 32.8$°/_{00}$). This mixing of saline water from the Bering Sea proper with less saline, but colder, shelf waters probably occurs over much of the region between St. Lawrence and St. Matthew islands. Alaskan Coastal Water, the least saline and the warmest water mass, is formed from a cold water of S $\approx$32$°/_{00}$ relict from the previous winter. It is seasonally diluted by Alaskan coastal runoff and warmed by solar heating. The degree of dilution varies with proximity to the Alaskan coast.

The base condition for all shelf waters in winter is isohaline and isothermal at the freezing point within each water column, but with salinity increasing westward across the system (from $\approx 32°/_{00}$ on the east side to $\approx 33°/_{00}$ on the west). The advent of runoff in spring leads to the two-layer regime observed in summer, with a less saline and hence less dense surface layer of 10 to 15 m thickness overlying the relatively uniform deeper water. Where salinity differences between the upper and lower layers remain relatively small, lower layer temperatures increase with time due to a downward heat flux, thus preventing a large accumulation of heat and high temperatures in the upper layer. On the other hand, where the salinity differences between layers are great, downward heat flux is inhibited by the stability, so that upper layer temperatures rise toward the climatic limit determined by surface exchange.

This stability-controlled thermal regime is probably a common feature of all Arctic shelf seas, and it has an important influence on ice distribution. Waters with very stable surface layers are centers of ice disintegration in spring, and the advection of these waters significantly influences summer ice distributions.

The seasonal variation in the fresh water component of the water masses is correlated with the discharge of the Yukon River, which serves as an index to the regional runoff. The fresh water additions are greatest in June and July, but fresh water continues to accumulate in the region over the summer, reaching a maximum in September. The heat content of the system parallels that of fresh water. In the fall about two months are required to flush the accumulated heat and fresh water from the region, so that significant ice formation typically does not occur until November/December.

Since the river discharge varies from year to year, and the fresh water accumulation is correlated with the runoff, salinity values also vary year to year. For example, the range of mean summer salinities of Anadyr Water at the present time is 32.8 to $33.2°/_{00}$. There is also evidence of a freshening of the system since the 1930s of about $0.2°/_{00}$ and, because of the stability-controlled temperature regime, an associated surface warming trend of about 1°C. However, because of the greater insulating effect of a fresher surface layer, the total heat content of the system varies inversely with the surface temperature, so that the average heat content has actually decreased over the same four-decade period.

Two anomalous events have been documented, during July 1967 and February 1968. These events are considered anomalous because for short periods (about 1 week) the regional flow regime reversed from northerly to southerly. Under these conditions water masses in Bering Strait assume somewhat altered T-S relations, in particular reflecting enhanced lateral mixing between Alaskan Coastal and Bering Shelf waters. However, because the whole region, including the southern Chukchi Sea, is so dominated by the water masses of southerly origin advected north by the normal flow, prolonged flow reversals (probably greater than 2 weeks) would be required to flush the system so that Arctic water could be observed in the strait. There is no evidence that such prolonged flow reversals ever occur.

Only sparse data are available from which to evaluate short-period (one day or less) local property variations. These indicate that there are numerous eddies, of horizontal dimensions 10 to 30 km, which are being advected north with the mean flow.

CURRENTS

The pattern of velocity shear in the Bering Strait section appears to remain relatively invariant regardless of the magnitude of transport. The swiftest flows (which can be more than 200 cm sec^{-1}) are always in the upper layers of the eastern channel, with a vertical shear such that the deeper speeds are reduced by about one-half. Immediately westward of the swift flow, there is a strong horizontal shear, so that speeds in the western channel are one-fourth to one-third those in the eastern channel and relatively uniform across the channel. This consistent pattern suggests the possibility of monitoring the transport through the strait with one or two fixed current meter installations, suitably calibrated.

Away from Bering Strait, the swiftest flow is adjacent to major promontories, notably near Point Hope north of Kotzebue Sound and in the channels east and west of St. Lawrence Island. Apart from these local effects, the measured currents are in general much slower and more variable in direction than in Bering Strait, though they agree on the average with the flow deduced from the water mass distributions. There is evidence of eddies in the system, some as small as a few km, with the possibility that a few of the larger ones may be semipermanent features, probably bathymetrically induced. For example, the latter may occur northwest of Shishmaref (in the lee of Cape Prince of Wales) and northeast of Cape Lisburne (also in the lee).

Prominent semidiurnal oscillations appear in the anchored current records from both north and south of Bering Strait, with net (peak-to-peak) amplitudes typically about 10 cm sec^{-1}. Diurnal oscillations are one-fourth to one-half as great. We conclude that these oscillations are primarily tidal. The tidal situation appears to be rather complex, with several different tide waves varying in their relative importance over the region. There appear to be at least three amphidromes: (1) a cyclonic diurnal system in Norton Sound, (2) a cyclonic system immediately south of St. Lawrence Island, and (3) a cyclonic semidiurnal system in the Gulf of Anadyr.

Other short-term variations in the recorded currents include (1) vertically incoherent local flow reversals, probably associated with the advection of thin baroclinic eddies past the meters; (2) quarter-diurnal and longer oscillations which also appear induced by eddies; and (3) fluctuations with characteristic times of 1 to 3 days, at least some of which may be forced by local wind oscillations.

Transport through the region has been documented by eleven detailed sections of current measurements across Bering Strait and ten sections located to the north and south which traverse the system. Transports varied from more than 2 Sv north to 0.2 Sv south. The transport was well correlated (correlation coefficient of 0.79) with surface atmospheric pressure deviations at Nome; the equation

$$\text{Transport (Sv)} = 1.58 + 0.08 \, (P_N - \bar{P})$$

will forecast the transport within about 0.5 Sv. The reason for the good correlation appears to be that the Nome pressure reflects to a high degree the regional pressure distribution. With high pressure at Nome there is normally low pressure north of Bering Strait, and the associated wind fields tend to cause a convergence to the south and a divergence to the north, enhancing north flow. Conversely, when the pressure is low at Nome, the regional winds cause a divergence in the northern Bering Sea.

The local winds further modify the transport, primarily through driving the surface layer, and the effect is about 0.5 Sv for each dyne cm^{-2} of wind stress applied. There is no evidence for an annual variation in Bering Strait transport. Rather, the transport can vary by a factor of two within about one week. Mean monthly and annual transports are close to 1.5 Sv north.

CHUKCHI SEA

The three water masses flowing through Bering Strait transit the Chukchi Sea en route to the Arctic Ocean. Immediately north of the strait they are reduced to two through combination of Anadyr and Bering Shelf waters to form Bering Sea Water. Alaskan Coastal Water maintains a separate identity and is added to by contributions of low-salinity water from Kotzebue Sound.

There is a wave-like pattern to the northward flow in the southern Chukchi Sea, at least in part due to bathymetric steering. The flow curves east toward Kotzebue Sound and then turns north and west, accelerating in the vicinity of Point Hope.

Off Point Hope the flow bifurcates. Most of the Bering Sea Water (salinity in the range 32.5 to 33°/$_{oo}$) goes northwest toward Herald Island and thence into the Arctic Ocean, following the Hope Submarine Valley. Most of the Alaskan Coastal Water turns northeast off Point Hope toward Point Barrow, moves along Barrow Canyon, and enters the Beaufort Sea of the Arctic Ocean. The net northward transport through the Chukchi increases by about 0.5 Sv from south to north. The Alaskan Coastal Water increases in volume due to additions from Kotzebue and possibly from some entrainment of resident Chukchi Sea water; it is semiisolated by the Alaskan shoreline on the east. In contrast, the Bering Sea Water actually decreases in identifiable volume, probably due to a greater mixing along its more exposed pathway across the central Chukchi Sea. The temperatures of both branches are modified en route, and we estimate vertical eddy conductivities of 0.2-2 cm^2 sec^{-1}.

In the south-central Chukchi Sea there is an apparent "dead spot" in the circulation, with Bering Sea Water bypassing it on the north. The positions of the dead spot and its border with the main current of Bering Sea Water shift by as much as a few 10s of kilometers. Bottom water in the dead spot is residual from winter, and oxygen utilization rates are about 9 ml/l/yr.

Intrusions of relatively dense water into the Chukchi come from two sources: (1) A cold East Siberian Sea bottom water enters from the west through Long Strait under appropriate atmospheric conditions; this water flows southeast along the Siberian coast, but rarely reaches past Cape Serdtse-Kamen; and (2) relatively saline water from the Arctic Ocean may enter from the north along the bottoms of both Herald and Barrow submarine canyons. For example, in Barrow Canyon we have measured pulses of Atlantic Water moving onto the shelf as fast as 40 cm sec^{-1} and of duration over two days (Mountain et al., in press).

NUMERICAL CONSIDERATIONS

The dynamics of the flow in the vicinity of Bering Strait is that of a northward barotropic pressure gradient force (down-sloping sea surface), balanced on the average by (a) 60% frictional retardation (shear stress) plus (b) 35% lateral advection of momentum. In the majority of cases the baroclinicity opposed the north flow but contributed less than 10% to the total force. Transport is highly correlated with sea surface slope. For an average summer transport of 1.7 Sv the slope is 2.4×10^{-6}

down to the north, and it was reversed (down to the south) for the one documented case of south flow. The slope is less away from the strait.

Lateral and vertical eddy coefficients were calculated where permitted by the data. Vertical coefficients (K_V) range from about 0.1 cm^2 sec^{-1} in regions with high stability to 10 cm^2 sec^{-1} in low stability regimes. Horizontal coefficients (K_H) are of order 10^6 cm^2 sec^{-1} and seem to increase eastward across the region between St. Lawrence Island and Bering Strait; i.e., lateral diffusion is largest close to the Alaskan coast.

A box model of the heat content of the region from St. Lawrence to north of Bering Strait showed that in July the mean surface heat flux (ignoring lateral fluxes) was larger than could be accounted for by local addition through the sea surface. This discrepancy points toward Norton Sound as a source of summer heat for the northern shelf. Three subregions showed local net heat losses. Two were associated with Alaskan Coastal Water, one along the coast from Nome to Port Clarence and the other along Cape Prince of Wales Shoal. The third was in the Strait of Anadyr. These deficits must be due to a lateral loss of heat to adjacent waters.

Oceanographic programs planned for the region in the future might consider investigating some of the following problems. The list is not meant to include all the important but little understood regional phenomena, but rather to focus attention on what we feel are problems whose solutions would add significant new increments of knowledge of general interest and application.

(1) An important oceanographic problem is the nature of major reversals in flow fields. Such changes are apparent in the Bering Strait transports and in near-bottom current records recently obtained from Barrow Canyon. Appropriate data would be long (more than one year) current records from strategic locations, and we suggest:

A. A few bottom-moored installations in Bering Strait. This may prove difficult, and due precaution must be taken to protect against possible ice scour. The record could probably serve as an index to the transport, with proper calibration, because of the relatively invariant cross-strait shear pattern.

B. Long-term moorings in Barrow and Herald canyons and Long Strait, the meters to include a temperature sensor.

(2) As descriptive oceanographic data become more frequent in time and space, eddies and eddy-like features are being recognized as common components of flow fields. Recently, small (15 km or less) baroclinic eddies have been identified in the Arctic Ocean pycnocline and appear to be quite common (Newton et al. 1974). These eddies are not generated

locally, but have been advected from some other area, with the Chukchi Sea a likely source. We have indicated the probable existence of numerous eddies in the region, and as these are constrained in their trajectories by the local bathymetry, the vicinity of Bering Strait seems well suited to economical experiments designed to elucidate the dynamics of eddies.

(3) The regional tidal wave behavior must be regarded as unknown. Some of the seemingly peculiar behavior we have found, for example, why there appears to be so much baroclinic decoupling in the vertical direction when the system is not strongly stratified and seems to behave barotropically much of the time, may be pertinent to the general behavior of shelf tides. A numerical tidal model of the region using Arctic Ocean tides from the north and Bering Sea tides from the south as boundary conditions might be possible, but undoubtedly further judiciously chosen measurements would also be required.

(4) The area south of St. Lawrence and north of St. Matthew islands is the least understood. Here Bering Shelf Water is formed, as is "cold center" water (Barnes and Thompson 1938) so important to the Bering Sea shelf to the south. Oxygen utilization rates in the near-bottom water appear to be extremely high, and the area is the winter home of an enormous marine mammal population (walrus). We do not as yet even know the general circulation obtaining here, and a systematic measuring program covering the region over an annual cycle is suggested.

REFERENCES

Aagaard, K. 1964
Features of the physical oceanography of the Chukchi Sea in the autumn. Univ. Wash. Dept. Oceanography, M.S. thesis, 41 pp.

Aagaard, K., and L. K. Coachman
Recent studies on Arctic currents. Proceedings of the Polar Oceans Conference, Montreal, in press.

Antonov, V. S. 1968
The nature of water and ice movement in the Arctic Ocean. *AANII. Trudy* vol. 285, pp. 154–82. (transl.)

Arsenyev, V. S. 1967
The currents and water masses of the Bering Sea. Moscow: Izdatels'tvo "Nauka" (transl.).

Barnes, C. A., and E. E. Collias 1958
Some considerations of oxygen utilization rates in Puget Sound. *J. Mar. Res.* 17:68–80.

Barnes, C. A., and T. G. Thompson 1938
Physical and chemical investigations in Bering Sea and portions of the north Pacific Ocean. Univ. Wash. Publ. Oceanog. vol. 3, no. 2, pp. 35–79.

Bloom, G. L. 1956
Current, temperature, tide, and ice growth measurements, eastern Bering Strait-Cape Prince of Wales 1953–1955. U.S.N. Electron. Lab., Res. Rept. 739. 24 pp.

Bloom, G. L. 1964
Water transport and temperature measurements in the eastern Bering Strait 1953–1958. *J. Geophys. Res.* 69:3335–54.

Coachman, L. K., and C. A. Barnes 1961
The contribution of Bering Sea water to the Arctic Ocean. *Arctic* 14(3):147–61.

Coachman, L. K., and C. A. Barnes 1962
Surface water in the Eurasian Basin of the Arctic Ocean. *Arctic* 15(4):251–77.

Coachman, L. K., and K. Aagaard 1966
On the water exchange through Bering Strait. *Limnol. and Oceanog.* 11(1):44–59.

Coachman, L. K., and D. A. Rankin 1968
Currents in Long Strait, Arctic Ocean. *Arctic* 21(1):27–38.

Coachman, L. K., and R. B. Tripp 1970
 Currents north of Bering Strait in winter. *Limnol. and Oceanog.* 15(4):625–32.
Countryman, K. A., and M. T. Bourkland 1968
 Oceanographic cruise summary, Bering Sea, June 1968. U.S.N. Oceanog. Off. Informal Rpt. IR 68–30.
Creager, J. S., and D. A. McManus 1964
 Notes on bottom sediments of the Chukchi Sea. U.S.C.G. Oceanog. Rpt. no. 1, pp. 23–24.
Creager, J. S., and D. A. McManus 1966
 Geology of the southeastern Chukchi Sea. Chap. 26, pp. 755–86, *in* Environment of the Cape Thompson Region, Alaska. U.S. Atomic Energy Commission, Div. of Technical Information.
Fedorova, A. P., and A. S. Yankina 1964
 The passage of Pacific Ocean water through the Bering Strait into the Chukchi Sea. *Deep-Sea Res. 11*:427–34 (transl.).
Fjeldstad, J. E. 1936
 Results of tidal observations. Norw. N. Polar Exped. *Maud*, 1918–1925, Sci. Res. vol. 4, no. 4, 88 pp.
Fleming, R. H., and D. Heggarty 1966
 Oceanography of the southeastern Chukchi Sea. Chap. 25, pp. 697–754 *in* Environment of the Cape Thompson Region, Alaska. U. S. Atomic Energy Comm., Div. of Tech. Information.
Fletcher, J. O. 1965
 The heat budget of the Arctic Basin and its relation to climate. RAND Corp. R-444-PR, 179 pp.
Goodman, J. R., J. H. Lincoln, T. G. Thompson, and F. A. Zeusler 1942
 Physical and chemical investigations: Bering Sea, Bering Strait, Chukchi Sea during the summers of 1937 and 1938. Univ. Wash. Publ. Oceanog. vol. 3, no. 2, pp. 105–169.
Gorbunov, Y. A. 1957
 On the water exchange between East Siberian and Chukchi seas through the Straits of Long. *Probl. Arktiki*, no. 1, pp. 35–40 (transl.).
Hufford, G. L., and D. M. Husby 1972
 Oceanographic survey of the Gulf of Anadyr, August 1970. U.S.C.G. Oceanog. Rpt. no. 52. 56 pp.
Husby, D. M., and G. L. Hufford 1971
 Oceanographic investigation of the northern Bering Sea and Bering Strait, 8–21 June 1969. U.S.C.G. Oceanog. Rpt. no. 42. 54 pp.
Ingham, M. C., B. A. Rutland, P. W. Barnes, G. E. Watson, G. I. Divoky, A. S. Naidu, G. D. Sharma, B. L. Wing, and J. C. Quast 1972
 Websec-70.An ecological survey in the eastern Chukchi Sea. September–October 1970. U.S.C.G. Oceanog. Rpt. no. 50, 206 pp.
Ketchum, B. H., and D. J. Keen 1955
 The accumulation of river water over the continental shelf between Cape Cod and Chesapeake Bay. *In* Marine Biology and Oceanography, Supplement to vol. 3 of *Deep-Sea Res.*, pp. 346–57.
Knebel, H. J. 1972
 Holocene sedimentary framework of the east-central Bering Sea conti-

nental shelf. Univ. Wash. Dept. Oceanogr. Ph.D. thesis. 132 pp. and appendices.

Leonov. A. K. 1960
Regional Oceanography, Part I. Leningrad: Gidrometizdat. 765 pp. (transl.).

Lesser, R. M., and G. L. Pickard 1950
Oceanographic cruise to the Bering and Chukchi seas, summer 1949–Part II–currents. U.S.N. Electron. Lab. Report 211, 15 pp.

Lewis, E. L., and R. A. Lake 1971
Sea ice and supercooled water. *J. Geophys. Res.* 76(24):5836–41.

Lisitsyn, A. P. 1966
Recent Sedimentation in the Bering Sea. Moscow: Izdatel'stvo "Nauka" (transl.).

Maksimov, I. V. 1945
Determining the relative volume of the annual flow of Pacific water into the Arctic Ocean through Bering Strait. *Probl. Arktiki*, no. 2, pp. 51–8 (transl.).

Marshunova, M. S. 1961
Osnovie zakonomernosti radiatsionnogo balansa podstilayushchei poverkhnosti i atmosfery v Arktike. *AANII Trudy* vol. 229. 53 pp.

Meilakh, I. G. 1958
K voprosu o proniknovenii vod severnogo ledovitogo okeana v Beringovo More. *Probl. Arktiki*, no. 3, pp. 35–40.

Mountain, D. G., L. K. Coachman, and K. Aagaard.
On the flow through Barrow Canyon. *J. Phys. Oceanog.* In press.

Neumann, G., and W. J. Pierson 1966
Principles of Physical Oceanography. Englewood Cliffs, N.J.: Prentice-Hall.

Newton, J. L., K. Aagaard, and L. K. Coachman 1974
Baroclinic eddies in the Arctic Ocean. *Deep-Sea Res.* 21(9):707–19.

Office of Climatology and Oceanographic Analysis Division 1961
Climatological and oceanographic atlas for mariners. vol. 2. N. Pacific Ocean. 165 pp.

Okubo, A., and R. V. Ozmidov 1970
Empirical dependence of the coefficient of horizontal turbulent diffusion in the ocean on the scale of the phenomenon in question. *Izv. AN/SSSR Fizika Atmos. i Okeana*, vol. 6, pp. 534–36 (transl.).

Ohtani, K. 1969
On the oceanographic structure and the ice formation on the continental shelf in the eastern Bering Sea. *Bull. Fac. Fish. Hokkaido Univ.* 20(2):94–117 (in Japanese).

Proudman, J. 1953
Dynamical Oceanography. London: Methuen.

Ratmanov, G. E. 1937a
K gidrologii beringova i chukotskogo morei. *Issledovaniya Morei SSSR* Vip. 25, 10–118.

Ratmanov, G. E. 1937b
K voprosu o vodoobmene cherez beringov proliv. *Issledovaniya Morei SSSR* Vip 25, 119–135.

Saur, J. F. T., J. P. Tully, and E. C. LaFond 1954
   Oceanographic cruise to the Bering and Chukchi seas, summer 1949. Part IV: Physical oceanographic studies: vol. 1. Descriptive report U.S.N. Electr. Lab. Report 416, vol. 1, 31 pp.

Shtokman, V. B. 1957
   Vliyanie vetra na techeniya v Beringovo Prolive, prichiny ikh bol'shikh skorostei i preobladayushego severnogo napravleniya. *Tr. Inst. Okeanol., Akad. Nauk S.S.S.R.*, 25:171–197.

Sverdrup, H. U. 1927
   Dynamic of tides on the North Siberian Shelf. *Geofys. Publ.*, vol. 4, no. 5, 75 pp.

Sverdrup, H. U. 1929
   The waters on the North Siberian Shelf. The Norweg. North Polar Exped. with the *Maud* 1918–1925. *Sci. Res.* vol. 4, no. 2, 131 pp.

Sverdrup, H. U., M. W. Johnson, and R. H. Fleming 1942
   The Oceans. New York: Prentice-Hall.

Tully, J. P., and F. G. Barber 1960
   An estuarine analogy in the sub-arctic Pacific Ocean. *J. Fish. Res. Bd. Canada* 17(1):92–112.

U.S. Navy Hydrographic Office 1954
   Oceanographic observations, *U.S.S. Burton Island*, 1950–1953. H.O. Publ. 618–c, 309 pp.

U.S. Navy Hydrographic Office 1958
   Oceanographic atlas of the polar seas. Part II. Arctic. H.O. Pub. no. 705.

INDEX

Accelerations: in Barrow Canyon, 123–24, 162; in Bering Strait, 75, 147–50; in other straits, 75–76, 160; near promontories, 75, 112, 141, 160, 162. See also Dynamical calculations

Alaskan Coastal Water: characteristics and modification in Chukchi, 129, 134–40, 144, 161–62; deep water salinities, 36; influence of fresh water, 36–41, 56; median salinities, 14, 158; temperature cycle, 43, 52, 54; temperature/salinity relationships, 39; thermal regime, 39–43; warming, 42–43, 63, 156, 163; winter base condition, 37, 159. See also Water masses

Amphidromic points, 90, 161

Anadyr Strait. See Strait of Anadyr

Anadyr Water: circulation, 23, 27, 33–34; continuity with Bering Sea, 21–27, 70, 72; influence of fresh water, 56–57, 159; median salinities, 14, 55–57, 158; minimum temperatures, 19; modification in Chukchi Sea, 114, 161; modification in Gulf of Anadyr, 23–27. See also Water masses

Anomalous events: possible incidences, 66–73, 108, 159; transports, 99–100, 102, 106–9; water mass distributions, 67–72, 117–18, 135, 160

Arctic Ocean: tide wave behavior, 91, 164; water intrusions in Chukchi Sea, 114, 123–27, 162; Bering Sea water, 111, 121, 134, 137, 139–40

Atmospheric disturbances, 94–95, 106

Atmospheric pressure: influences on sea levels, 105–6, 151–52; correlations with transports, 102–10, 144–45, 161

Baroclinic effects, 80, 83, 96–97, 150, 161–62, 164

Barrow Canyon: Arctic Ocean water, 114, 127; flow, 106, 123–24

Bering, Capt. Vitus, 3

Bering Sea, 15, 23, 27, 90, 164

Bering Sea Water: characteristics, 127; formation, 114, 128, 161; intrusion in Gulf of Anadyr, 24–26, 158; modifications in Chukchi, 128–34, 144. See also Water masses

Bering Shelf Water: formation south of St. Lawrence, 27–35, 158, 164; influence of fresh water, 56–57; layering and mixing, 31–33, 67, 160; median salinities, 14, 55–57, 158; modification in Chukchi Sea, 114. See also Water masses

Bering Slope Current, 27

Bering Strait: anomalous events, 67–72, 101; currents, 75–78, 107; dynamics of flow, 146–52, 162–63; horizontal shear, 76, 78, 140, 160; long-term trends, 54–58, 159; mean hydrographic conditions, 47–50; penetration of Chukchi Sea water, 117–18, 160; tidal wave behavior, 90–91; transports, 99–110, 161
—annual cycle: temperature, 47, 52–54; salinity, 47–52

Bottom: friction, 97–98, 147–50; influence on currents, 27, 133. See also Topography

Brown Bear, 8, 36, 40–41, 74–75, 112, 118, 135–36

Canyons. See Submarine canyons

Capes: Billings, 91; Chukotskiy, 71; Dezhneva, 6, 69, 74, 113, 117; Franklin, 111, 127, 140; Icy, 114, 119, 123, 135–36, 138; influence on currents, 75–76, 141, 160; influence on wind, 5–6; Krigugan, 95, 98; Lisburne, 74–75, 95, 100, 103, 114, 119–20, 125, 129–31, 135, 137–38, 142, 149, 160; Navarin, 21, 24, 26–27, 79; Northeast, 95–98, 103, 149; Peek, 11; Prince of Wales, 4, 6, 11, 68, 76, 112, 160; Rodney, 100; Romanov, 100; Schmidt, 91, 100, 103–4, 108; Serdze-Kamen, 113, 132, 162; Uelen, 103

Cedarwood, 8, 14, 16

Channels: east of St. Lawrence, 4; Hope submarine valley, 4; into Norton Sound, 4, 44; in Strait of Anadyr, 4

Chelan, 8, 60

Chukchi Sea: circulation, 125, 140–43, 161–62; exchange through Long Strait, 126, 132, 144–45, 162; fronts, 120, 126, 129, 132; gyre, 113, 125–26, 133, 138, 162; penetration of Arctic Ocean water, 123–24, 162; tides, 91; transports, 141–44, 162; water masses, 112–40, 160

169

Cold center, cold spot, 28, 30, 32, 34, 71–72, 158, 164
Conductivity. See Eddy coefficients
Convection, 29, 34–37, 45
Currents: Barrow Canyon, 106, 123–24; Bering Strait, 66, 76, 78, 109, 140, 160; bifurcation off Pt. Hope, 112, 135, 162; Chukchi Sea, schematic, 142–43; coherence, 81–82, 85, 88, 92, 161; convergence near promontories, 75–76, 160; diurnal variations, 93, 160; Gulf of Anadyr, 26–27, 70–71; gyres, 76, 138; horizontal shear, 76, 99, 140; influence of bottom topography, 33–34, 111–12, 125, 133; longer period fluctuations, 82–83, 93–95, 161; regional surveys, 74–77; semidiurnal variations, 79, 84–91, 160; spectral characteristics, 83–85, 91–95; variations in straits, 76, 97; vertical shear, 80, 96, 98, 160. See also Tidal currents

Drift bottle studies, 44
Dynamical calculations, 146–52

East Siberian Sea: as a source of waters for Chukchi Sea, 113, 120, 122; winter conditions, 126
Eddies: baroclinic, 81–82, 92–93, 161, 163–64; in the lee of C. Prince of Wales, 116, 160; in the lee of Pt. Hope–Cape Lisburne, 125, 136, 138, 160; north of St. Lawrence, 60–62, 76; size estimates, 62, 93; transient in Bering Sea, 62, 92–93, 160
Eddy coefficients: calculations, 33, 139, 152–54; lateral, 33, 154, 156, 163
—vertical, 42, 98, 139–40, 162–63; correlation with stability, 42, 153–54, 163
Ekman transport, 101
Embayments: influence on heat content, 155–56, 163; Norton Sound, 43–44, 163

Fall cooling. See Heat; Temperature
Flow reversals. See Anomalous events
Freezing: influence on shelf hydrography, 29, 37, 43, 138, 159; relation of temperatures to, 38, 45–46. See also Supercooling
Fresh water: accumulation, 50–53, 102, 159; connection with Alaskan Coastal Water, 36; correlation with salinities, 55–57, 113; correlation with transport, 63–64; equivalent area, 51–53, 67; flushing, 51, 54, 159; relationship with stability, 37–39, 42; seasonal cycle, 37, 44–54, 70, 159. See also Runoff
Friction: in dynamical calculations, 148–50, 162; influence on tidal currents, 97–98
Fronts. See Chukchi Sea; Water masses

Geostrophic balance, 150
Gradients, 36, 60, 112
Gulf of Anadyr: circulation, 26–27, 33–34, 72; current measurements, 26–27, 70–71; layering and mixing, 31–32, 158; tides, 90, 161; topography, 4

—water masses, 21–35, 158; distributions during anomalous event, 70–72
Gyres. See Eddies

Heat: budget, 154–56; content, 52, 159, 163; correlation with fresh water (runoff), 43, 54, 59, 159; seasonal cycle, 45, 54, 159; vertical fluxes, 42, 129, 136–40, 153–55, 159
—surface exchange, 112, 155–56, 159, 163; relationship with lateral eddy conductivity, 156, 163

Ice: centers for disintegration, 43, 159; formation, 34, 43, 159; influence on hydrographic conditions, 43, 136. See also Freezing
Inertial: accelerations, 147–50; period (frequency), 79, 84, 87, 89
Investigations, 6–10
Islands: Big Diomede (Ratmanov), 4, 11; Fairway Rock, 4, 11, 76, 90; Herald, 74, 111, 127, 133, 140, 144, 162; influence on flow field, 4; King, 4, 100; Little Diomede, 4, 11; Pribilof, 43; Ratmanov, see Big Diomede; St. Lawrence, 4, 18–20, 27, 34–36, 42, 70–72, 76, 90, 95, 97, 100, 102, 149, 158, 160–61, 163–64; St. Matthew, 34–35, 158, 164; Sledge, 4, 90; Wrangel, 120, 122, 126, 132, 144–45
Isopycnal: depth of surfaces, 97, 99; slopes, 33, 81

Kara Sea, submarine canyon in, 44
Kotzebue Sound, 4, 5, 10, 64, 103, 111–12, 125, 133, 140, 160, 162; influence on Alaskan Coastal Water, 115–16, 119–20, 134–35, 144, 161
Kolyuchin Bay, 126
Kresta Bay, 26, 71, 79

Lateral mixing. See Mixing
Layering, 31, 129, 134; development of two-layer structure, 47–49
Long Strait: currents, 94, 144–45, 163; water masses, 129, 132, 162

Maud, 112, 127, 140, 157
Meanders. See Eddies
Mixing: in relation to isopycnal slopes, 33, 132; lateral, 16, 17, 20, 21, 23, 27, 31, 33, 63, 131, 134, 144, 153–54, 156, 158, 160, 162–63; vertical, 33, 36, 132, 134
Momentum balance. See Dynamical calculations

Nome: tides, 91
—atmospheric pressure, 103–6; correlation with transport, 104–5, 108–9; index to regional winds, 107
Northland, 8, 44, 57
Northwind, 6–9, 17, 21, 24–29, 34–35, 44–45, 47, 52, 63, 70–72, 79, 107, 111–12, 114, 118, 127–28, 131, 146–47, 149
Norton Sound: circulation, 44; influence on

Alaskan Coastal Water, 39, 43–44, 155, 163; tides, 90, 161; topography, 4

Oceanographic surveys: design criteria, 6; summary of regional, 8–9
Oshoro Maru, 6, 9, 17, 21, 28, 74, 114, 118, 127, 129, 132–33, 135–36, 139–41
Oxygen: deficient bottom water, 113; utilization rates, 35, 113, 126, 162, 164
—percent saturation: in Chukchi Sea, 113, 124–26; in Gulf of Anadyr, 34–35

Point Barrow, 74, 144–45, 162
Point Hope, 75, 100, 111–12, 114, 119, 125, 128, 133–35, 160, 162
Point Spencer, 91
Port Clarence, 4, 156, 163
Pressure: longitudinal gradients, 106, 148–50, 161–62. *See also* Atmospheric pressure; Dynamical calculations.
Providence Bay, 103

Radiation, 42, 45
Rivers: Anadyr, 5, 38; Indigirka, 113; Kobuk, 5, 36; Kolyma, 113, 114; Kuskokwim, 39; Noatak, 5, 36. *See also* Fresh water
—Yukon, 5, 36–38, 40–41, 44, 49, 52, 55, 56, 102; index of runoff, 41, 52–53, 102, 159
Rossby number, 150
Ruby, Alaska, 37, 40, 52, 53, 55, 56, 102
Runoff: effect on salinities, 49–50, 55–57, 159; influence on sea surface slope, 102; seasonal cycle, 37, 39, 44–45, 159; year-to-year variations, 41, 56–57, 159. *See also* Fresh water

Salinity: definition of water masses, 13; distribution in Bering Strait, 11, 47–50, 67–69; distribution on Bering Sea shelf, 21–22, 36–37; distribution in Chukchi Sea, 112, 118–24; gradients, 36, 49, 60, 112, 120, 126, 137, 159; long-term variations, 55–57; seasonal cycle, 47–52; short-term variations, 60–66; variations with runoff, 55–57, 159; vertical fluxes, 43, 136. *See also* Water masses
Sea level: correlation with atmospheric pressure, 106; correlation with transport, 106, 150, 162; influence of regional winds, 102–6, 151; influence of runoff, 102; surface slope, 102, 147–48, 150–51, 162–63
Seiches, 95
Shishmaref, 64, 100, 114, 116, 147, 160
Shoals: Cape Prince of Wales, 4, 111–12, 140, 156, 163; Herald, 111–12, 119, 121, 134–36; Pt. Hope, 4, 111
Siberian Coastal Water. *See* Water masses
Stability: correlation with temperature, 42, 159; effect on mixing, 42; relationship with fresh water, 38
Staten Island, 6, 9, 10, 17, 36, 38–41, 44–45, 47, 52, 70, 74, 77, 95, 107, 146–47, 149
Steric level, 151–52
Straits: Anadyr, 4, 18, 21, 33, 42, 70–72, 75–76, 95–98, 107, 109, 156, 160, 163; current measurements, 71, 95–97, 107, 109; distribution of water masses, 34; horizontal shear, 76, 140; Long, 94, 111, 114, 120, 122, 126, 129, 132; St. Lawrence/Alaska, 18, 75–76, 160. *See also* Bering Strait
Submarine canyons: Barrow, 106, 111, 114, 122–23, 125, 127, 150, 162–63; Herald, 111, 119, 122, 125, 127, 129–32, 134, 140, 162–63; Hope, 4, 162; influence on flow field, 125; in Norton Sound, 44; role as sunken estuaries, 44, 127; Svataya Anna, 44
Supercooling, 46

Temperature: of cores, 19–20, 130, 132–33, 135, 137, 153, 162; correlation with fresh water (runoff), 41, 52, 58–59, 159; correlation with salinity, see Water masses; correlation with stability, 39–42; distribution in Bering Strait, 11, 47–50, 54, 67–69; distribution on Bering Sea shelf, 19, 21–22, 39; distribution in Chukchi Sea, 112, 118–24, 131; freezing point, 38, 45–46; intermediate maximum in Arctic Ocean, 139–40; mean values in Bering Strait, 47–50, 54, 57–59; seasonal cycle, 43, 45–47, 52–54, 63; short-term variations, 60–66; year-to-year variation, 58–59
Thompson, T. G., 57, 157
Thompson, 6, 9, 24, 26, 63, 146–47, 149
Tidal current: diurnal, 74, 91, 160; ellipses, 85–88, 93; reduction with depth, 96–97; semidiurnal, 74, 84–89, 91, 96–98, 160; shallow water constituents, 92. *See also* Tides
Tides: amphidromic points, 90, 161; cotidal lines, 90–91; diurnal, 91; ranges, 76, 89–90; reference stations, 89–91; regional variations, 90–91; wave behavior, 88, 90–91, 95–98, 161, 164
Topography: influence on currents, 33–34, 111–12, 125, 133, 158, 161. *See also* Shoals; Submarine canyons
—regional: Bering Sea shelf, 4–5; Chukchi Sea, 4, 111
Transports: annual cycle, 107–10, 152; in Chukchi Sea, 141–44, 162; correlation with atmospheric pressure, 103–10, 161; correlation with surface slope, 150–51, 162; correlation with wind, 99–102, 105–6, 109, 161; forecasting, 105, 161; of fresh water, 52–53; reversals, see Anomalous events
T-S relationships: basic regional characteristics, 11–17; continuity through Bering Strait, 19–20; water mass classification, 13–17. *See also* Water masses

Wales: tides, 91; winds, 6, 95
Water masses: boundaries, 20, 64; median salinities, 14, 55, 114, 121; mixing between, 23, 27–28, 71, 128, 133, 158, 162; position and paths of cores, 20, 129; ratios in Bering Strait, 66–67; variations associated with transport, 63–66

—modifications: in Chukchi Sea, 127–40; in Gulf of Anadyr, 21–35, 70–72
—north of Bering Strait: Alaskan Coastal, 112, 114–16, 119, 125, 134–40, 144, 161–62; Bering Sea, 114–15, 121, 125, 127–34, 144–45, 161–62; East Siberian Sea, 126–29, 132, 142, 162; resident Chukchi Sea bottom, 117, 125–27, 129–31, 136, 142, 162; Siberian Coastal, 69, 112–14, 116–18, 120, 122, 125, 128–29, 132, 162
—south of Bering Strait: Alaskan Coastal, 13, 36–44, 156, 158–60; Anadyr, 13, 21–27, 158; Bering Shelf, 13, 27–35, 158, 160, 164

Winds: correlation with transport, 99–102, 105–6, 161–62; influence on currents, 95, 112, 144–45, 161; influence on ice motion, 73; stress, 99